GENDER DIFFERENCES AT CRITICAL TRANSITIONS IN THE CAREERS OF SCIENCE, ENGINEERING, AND MATHEMATICS FACULTY

Committee on Gender Differences in Careers of Science, Engineering, and Mathematics Faculty

Committee on Women in Science, Engineering, and Medicine

Policy and Global Affairs

Committee on National Statistics

Division of Behavioral and Social Sciences and Education

NATIONAL RESEARCH COUNCIL
OF THE NATIONAL ACADEMIES

THE NATIONAL ACADEMIES PRESS
Washington, D.C.
www.nap.edu

THE NATIONAL ACADEMIES PRESS   500 Fifth Street, N.W.   Washington, DC 20001

NOTICE: The project that is the subject of this report was approved by the Governing Board of the National Research Council, whose members are drawn from the councils of the National Academy of Sciences, the National Academy of Engineering, and the Institute of Medicine. The members of the committee responsible for the report were chosen for their special competences and with regard for appropriate balance.

This project was supported by the National Science Foundation, Grant No. 0336796. Any opinions, findings, conclusions, or recommendations expressed in this publication are those of the author(s) and do not necessarily reflect the views of the organizations or agencies that provided support for the project.

Library of Congress Cataloging-in-Publication Data

Gender differences at critical transitions in the careers of science, engineering, and mathematics faculty / Committee on Gender Differences in Careers of Science, Engineering, and Mathematics Faculty, Committee on Women in Science, Engineering, and Medicine [of] Policy and Global Affairs [and] Committee on National Statistics, Division of Behavioral and Social Sciences and Education, the National Research Council of the National Academies.
    p. cm.
Includes bibliographical references and index.
ISBN-13: 978-0-309-11463-9 (hardcover)
ISBN-10: 0-309-11463-2 (hardcover)
ISBN-13: 978-0-309-11464-6 (pdf)
ISBN-10: 0-309-11464-0 (pdf)
  1. Universities and colleges—Faculty—Employment—Sex differences—United States—Statistics. 2. Sex discrimination in employment—United States. 3. Sex discrimination in higher education—United States. 4. Women in science—United States. 5. Women in technology—United States. 6. Women in mathematics—United States. 7. Educational surveys—United States. I. National Research Council (U.S.). Committee on Gender Differences in Careers of Science, Engineering, and Mathematics Faculty. II. National Research Council (U.S.). Committee on Women in Science, Engineering, and Medicine. III. National Research Council (U.S.). Committee on National Statistics.
  Q148.G46 2009
  331.4'133—dc22
                                          2009037397

Additional copies of this report are available from the National Academies Press, 500 Fifth Street, N.W., Lockbox 285, Washington, DC 20055; (800) 624-6242 or (202) 334-3313 (in the Washington metropolitan area); Internet, http://www.nap.edu.

THE NATIONAL ACADEMIES
Advisers to the Nation on Science, Engineering, and Medicine

The **National Academy of Sciences** is a private, nonprofit, self-perpetuating society of distinguished scholars engaged in scientific and engineering research, dedicated to the furtherance of science and technology and to their use for the general welfare. Upon the authority of the charter granted to it by the Congress in 1863, the Academy has a mandate that requires it to advise the federal government on scientific and technical matters. Dr. Ralph J. Cicerone is president of the National Academy of Sciences.

The **National Academy of Engineering** was established in 1964, under the charter of the National Academy of Sciences, as a parallel organization of outstanding engineers. It is autonomous in its administration and in the selection of its members, sharing with the National Academy of Sciences the responsibility for advising the federal government. The National Academy of Engineering also sponsors engineering programs aimed at meeting national needs, encourages education and research, and recognizes the superior achievements of engineers. Dr. Charles M. Vest is president of the National Academy of Engineering.

The **Institute of Medicine** was established in 1970 by the National Academy of Sciences to secure the services of eminent members of appropriate professions in the examination of policy matters pertaining to the health of the public. The Institute acts under the responsibility given to the National Academy of Sciences by its congressional charter to be an adviser to the federal government and, upon its own initiative, to identify issues of medical care, research, and education. Dr. Harvey V. Fineberg is president of the Institute of Medicine.

The **National Research Council** was organized by the National Academy of Sciences in 1916 to associate the broad community of science and technology with the Academy's purposes of furthering knowledge and advising the federal government. Functioning in accordance with general policies determined by the Academy, the Council has become the principal operating agency of both the National Academy of Sciences and the National Academy of Engineering in providing services to the government, the public, and the scientific and engineering communities. The Council is administered jointly by both Academies and the Institute of Medicine. Dr. Ralph J. Cicerone and Dr. Charles M. Vest are chair and vice chair, respectively, of the National Research Council.

www.national-academies.org

COMMITTEE ON GENDER DIFFERENCES IN CAREERS OF SCIENCE, ENGINEERING, AND MATHEMATICS FACULTY[1]

Claude R. Canizares, *Co-Chair*, Vice President for Research and Associate Provost and Bruno Rossi Professor of Experimental Physics, Massachusetts Institute of Technology

Sally E. Shaywitz, *Co-Chair*, Audrey G. Ratner Professor in Learning Development and Co-Director, Yale Center for Dyslexia and Creativity, School of Medicine, Yale University

Linda Abriola, Dean of Engineering, Tufts University

Jane Buikstra, Regents' Professor, Arizona State University School of Human Evolution and Social Change

Alicia Carriquiry, Professor of Statistics, Iowa State University

Ronald Ehrenberg, Director, Cornell Higher Education Research Institute and Irving M. Ives Professor of Industrial and Labor Relations and Economics, Cornell University

Joan Girgus, Professor of Psychology and Special Assistant to the Dean of the Faculty, Princeton University

Arleen Leibowitz, Professor of Public Policy, School of Public Affairs, University of California at Los Angeles

Thomas N. Taylor, Distinguished Professor, Department of Ecology and Evolutionary Biology and Biodiversity Research Institute, University of Kansas

Lilian Wu, Program Executive, Global University Relations, IBM Technology Strategy and Innovation

Study Staff

Catherine Didion, Study Director (from September 1, 2007)

Peter Henderson, Study Director (from October 15, 2005 until August 31, 2007)

Jong-on Hahm, Study Director (through October 14, 2005)

Constance F. Citro, Director, Committee on National Statistics

Michael L. Cohen, Senior Program Officer

John Sislin, Program Officer

Elizabeth Briggs, Senior Program Associate

Elizabeth Scott, Project Assistant

Jessica Buono, Research Associate

Jacqueline Martin, Senior Program Assistant

---

[1] Cathleen Synge Morawetz, Professor Emerita, the Courant Institute of Mathematical Sciences, New York University and Yu Xie, Frederick G. L. Huetwell Professor of Sociology, University of Michigan resigned their committee appointments in 2004.

iv

COMMITTEE ON WOMEN IN SCIENCE, ENGINEERING, AND MEDICINE 2007-2009

Lilian Shiao-Yen Wu, *Chair,* Program Executive, Global University Relations, IBM Technology Strategy and Innovation

Alice M. Agogino, Roscoe and Elizabeth Hughes Professor of Mechanical Engineering, University of California, Berkeley

Florence B. Bonner, Senior Vice President for Research and Compliance, Howard University

Allan Fisher, Senior Vice President, Product Strategy and Development, Laureate Higher Education Group

June E. Osborn, President Emerita, Josiah Macy, Jr. Foundation

Vivian Pinn, Associate Director for Research on Women's Health and Director, Office of Research on Women's Health, National Institutes of Health

Pardis Sabeti, Assistant Professor, Department of Organismic and Evolutionary Biology, Harvard University

Lydia Villa-Komaroff, Board Member and Chief Scientific Officer, CytonomeST, LLC

Warren Washington, Senior Scientist & Section Head, National Center for Atmospheric Research

Susan Wessler, University of Georgia Foundation Chair in Biology, Department of Plant Biology, University of Georgia

Staff

Catherine Didion, Director (from March 1, 2007)
Peter Henderson, Director (October 15, 2005-March 1, 2007)
Jong-on Hahm, Director (until October 14, 2005)
John Sislin, Program Officer
Elizabeth Briggs, Senior Program Associate
Jessica Buono, Research Associate
Jacqueline Martin, Senior Program Assistant

Preface

Difficult tasks are often very simply stated. This committee was asked by Congress to "conduct a study to assess gender differences in the careers of science, engineering, and mathematics (SEM) faculty, focusing on four-year institutions of higher education that award bachelor's and graduate degrees. The study will build on the National Academies' previous work and examine issues such as faculty hiring, promotion, tenure, and allocation of institutional resources including (but not limited to) laboratory space." That such an assessment would be daunting was well understood by the committee. The importance of the study provided more than ample motivation to keep the committee engaged and focused on crafting an objective report that would advance our knowledge on the status of women academics in science and engineering at the nation's top universities.

To address its charge, the committee drew on a large number of scholarly studies, survey data collected by federal agencies and professional societies among others, self-assessments conducted by universities—as well as a number of experts brought in to meet with the committee. After reviewing the above information, the committee determined to conduct two comprehensive surveys. These surveys were sent to the major research universities across the United States during 2004-2005. The surveys focused on biology, chemistry, civil and electrical engineering, mathematics, and physics. One focused on almost 500 departments in these disciplines, and the other was sent to more than 1,800 faculty. These surveys bring much needed additional information to the table. The survey of departments collected information on departmental characteristics, hiring, tenure and promotion decisions, and related policies. The survey of faculty focused on demographic characteristics, employment history, and institutional resources received. The committee was delighted with the response to the surveys. The departmental survey had about an 85 percent response rate, and the faculty survey had about a 77 percent response rate. The committee extends its thanks to everyone who filled out the questionnaires, which were undoubtedly time consuming. Respondents were very open with their information, as they were promised confidentiality. While the data must remain restricted to maintain that confidentiality, we believe these data could be used in further studies for the benefit of the scientific community without violating the confidentiality of respondents.

A related point is that while the committee examined a tremendous amount of information, a comprehensive and conclusive assessment of faculty careers remains in the future. The committee has done all it can given its resources to advance our understanding of this important issue, but additional research and study remain. If it could, this committee would have continued expanding, refining, and enhancing its analysis. The committee trusts that others will be encouraged to pursue further some of the avenues that the committee has started down and to answer some of the questions that arose in this report, drawing on their own innovative approaches to examining the trajectory of academic careers of men and women.

Claude R. Canizares
Co-Chair
Sally E. Shaywitz
Co-Chair

Acknowledgments

This report has been reviewed in draft form by individuals chosen for their diverse perspectives and technical expertise, in accordance with procedures approved by the National Academies' Report Review Committee. The purpose of this independent review is to provide candid and critical comments that will assist the institution in making its published report as sound as possible and to ensure that the report meets institutional standards for objectivity, evidence, and responsiveness to the study charge. The review comments and draft manuscript remain confidential to protect the integrity of the process.

We wish to thank the following individuals for their review of this report: Robert Birgeneau, University of California, Berkeley; Claudia Goldin, Harvard University; Susan Graham, University of California, Berkeley; Jo Handelsman, University of Wisconsin; Maria Klawe, Harvey Mudd College; J. Scott Long, Indiana University; Colm O'Muircheartaigh, University of Chicago; Barbara Reskin, University of Washington; Johanna Levelt Sengers, National Institute of Standards and Technology; and Richard Zare, Stanford University.

Although the reviewers listed above have provided many constructive comments and suggestions, they were not asked to endorse the conclusions or recommendations, nor did they see the final draft of the report before its release. The review of this report was overseen by Stephen Fienberg, Carnegie Mellon University and Mildred Dresselhaus, Massachusetts Institute of Technology. Appointed by the National Academies, they were responsible for making certain that an independent examination of this report was carried out in accordance with institutional procedures and that all review comments were carefully considered. Responsibility for the final content of this report rests entirely with the authoring committee and the institution.

We would like to thank all the faculty and departments who took the time to complete our surveys. We greatly appreciate the effort, and the report could not have been a success without their help. In addition, the committee would like to thank staff of the project for their assistance, including Charlotte Kuh, deputy executive director of the Policy and Global Affairs Division; Catherine Didion, the current director of the Committee on Women in Science, Engineering and Medicine (CWSEM); Michael Cohen, senior program officer of the Committee on National Statistics; Marilyn Baker, director for reports and communications for the Policy and Global Affairs Division; Jong-on Hahm, who managed the project as the former director of CWSEM; Peter Henderson, who took over as study director and interim director of CWSEM; John Sislin, program officer; Jim Voytuk, senior program officer and George Reinhart, former senior program officer for their assistance with data and surveys; Elizabeth Briggs; Jessica Buono; Jacqueline Martin; Amber Carrier; Melissa McCartney; Norman Bradburn, who consulted on the surveys and data; John Tsapogas for assistance with Survey of Doctorate Recipients data; Dan Heffron for assistance with National Survey of Postsecondary Faculty data; and Rachel Ivie, Roman Czujko, and everyone at the American Institute of Physics, who implemented the surveys.

Contents

List of Tables and Figures

TABLES

FIGURES

Summary

The 1999 report, *A Study on the Status of Women Faculty in Science at MIT*, created a new level of awareness of the special challenges faced by female faculty in the sciences. Although not the first examination of the treatment of female faculty, this report marked an important historical moment, igniting interest in the difficulties experienced by many women, particularly those at the higher levels of academia. Since the release of the Massachusetts Institute of Technology report, many other institutions have studied equity issues regarding their faculty, and several have publicly pledged to use their resources to correct identified disparities. Although academic departments, institutions, professional societies, and others have paid more attention to the topic in the past 10 years, some experts are concerned that remedial actions have approached a plateau.

Unquestionably, women's participation in academic science and engineering (S&E) has increased over the past few decades. In the 10 years prior to the start of this study, the number of women receiving Ph.D.s in science and engineering increased from 31.7 percent (in 1996) to 37.7 percent (in 2005). The percentage of women among doctoral scientists and engineers employed full-time, while still small, rose from 17 percent in 1995 to 22 percent in 2003. However, women continued to be underrepresented among academic faculty relative to the number receiving S&E degrees. In 2003, women comprised between 18 and 45 percent of assistant professors in S&E and between 6 and 29 percent of associate and full professors.

In 2002, Senator Ron Wyden (D-Oregon) of the Subcommittee on Science, Technology and Space of the U.S. Senate Committee on Commerce, Science and Transportation, convened three hearings on the subject of women studying and working in science, mathematics, and engineering. Soon after, Congress directed the National Science Foundation (NSF) to contract with the National Academies for a study assessing gender differences in the careers of science and engineering faculty, based on both existing and new data. The study committee was given the following charge:

> Assess gender differences in the careers of science, engineering, and mathematics (SEM) faculty, focusing on four-year institutions of higher education that award bachelor's and graduate degrees. The study will build on the Academy's previous work and examine issues such as faculty hiring, promotion, tenure, and allocation of institutional resources including (but not limited to) laboratory space.

The committee interpreted its charge to imply three tasks: (1) update earlier analyses, (2) identify and assess current gender differences, and (3) recommend methods for expanding knowledge about gender in academic careers in science and engineering. It developed a series of guiding research questions in three key areas to organize its investigation: (1) academic hiring, (2) institutional resources and climate, and (3) tenure and promotion.

The committee also limited its exploration of science and engineering to the natural sciences and engineering, defined here as the physical sciences (including astronomy, chemistry, and physics); earth, atmospheric, and ocean sciences; mathematics and computer science; biological and agricultural sciences; and engineering (in all its forms).

FACULTY AND DEPARTMENTAL SURVEYS

Recognizing at the outset the need for new data, the committee conducted two national surveys in 2004 and 2005 of faculty and academic departments in six science and engineering disciplines: biology, chemistry, civil engineering, electrical engineering, mathematics, and physics. The first survey of almost 500 departments focused on hiring, tenure, and promotion processes, while the second survey gathered career-related information from more than 1,800 faculty. Together the surveys addressed departmental characteristics, hiring, tenure, promotion, faculty demographics, employment experiences, and types of institutional support received. In addition to results from the surveys, the committee heard expert testimony, examined data from NSF, the National Center for Education Statistics (NCES), and professional societies, and reviewed the results of individual university studies and research publications.

As it would be impossible to survey all "science, engineering, and mathematics (SEM) faculty at four-year institutions of higher education," the committee limited the scope of the surveys in four important ways. These limitations must be kept in mind in the interpretation of the survey results:

1.   The data present a snapshot in time (2004 and 2005), not a longitudinal view.

2.   Six disciplines are examined: biology, chemistry, civil engineering, electrical engineering, mathematics, and physics.

3.   Institutions are limited to major research universities, referred to as Research I or research-intensive (RI) institutions.

4.   Only full-time, regularly appointed professorial faculty who are either tenure eligible or tenured are included.

In other words, except in its review of historical data and existing research, the report does not examine gender differences outside of the six disciplines covered in the surveys or at institutions other than RI institutions. It also does not examine the careers of instructors, lecturers, postdocs, adjunct faculty, clinical faculty, or research faculty, who may experience very different career paths.

Many of the "whys" of the findings included here are buried in factors that the committee was unable to explore. We do not know, for example, what happens to the significant percentage of female Ph.D.s in science and engineering who do not apply for regular faculty positions at RI institutions, or what happens to women faculty members who are hired and subsequently leave the university. And we know little about female full professors and what gender differences might exist at this stage of their careers.

We do know that there are many unexplored factors that play a significant role in women's academic careers, including the constraints of dual careers; access to quality child care; individuals' perceptions regarding professional recognition and career satisfaction; and other quality-of-life issues. In particular, the report does not explore the impact of children and family obligations (including elder care) or the duration of postdoctoral positions on women's willingness to pursue faculty positions in RI institutions.

COMPARISONS TO OTHER NATIONAL ACADEMIES' REPORTS

This report does not exist in isolation. The committee has benefited greatly from three other National Academies' reports on women in academic science and engineering. In 2001 the Committee on Women in Science and Engineering (CWSE) published *From Scarcity to Visibility: Gender Differences in the Careers of Doctoral Scientists and Engineers*, a statistical analysis of the career progression of matched cohorts of men and women Ph.D.s from 1973 to 1995. The 2005 CWSE report, *To Recruit and Advance: Women Students and Faculty in U.S. Science and Engineering,* identifies the strategies that higher education institutions have employed to achieve gender inclusiveness, based on case studies of four successful universities.

A third report, *Beyond Bias and Barriers: Fulfilling the Potential of Women in Academic Science and Engineering*, was released in 2006 under the aegis of the Committee on Science, Engineering, and Public Policy (COSEPUP). The study committee was charged to "review and assess the research on sex and gender issues in science and engineering, including innate differences in cognition, implicit bias, and faculty diversity" and "provide recommendations . . . on the best ways to maximize the potential of women science and engineering researchers." The committee considered all fields of science and engineering (including the social sciences) in a broad range of academic institutions, relying primarily on existing data and the experience and expertise of committee members. Its report provides broad policy recommendations for changes at higher education institutions.

In contrast, the current report examines new information on the career patterns of men and women faculty at RI institutions—with particular focus on key transition points that are under the control of the institutions. The findings and recommendations here are based primarily on the data from our two surveys, which were not available to the COSEPUP committee.

Like the COSEPUP committee, this committee found evidence of the overall loss of women's participation in academia. That loss is most apparent in the smaller fraction of women who apply for faculty positions and in the attrition of women assistant professors before tenure consideration. Unfortunately, our surveys do not shed light on why women fail to apply for faculty positions or why they may leave academia between these critical transition points—underscoring the fact that our work is not done.

Our survey findings do indicate that, at many critical transition points in their academic careers (e.g., hiring for tenure-track and tenured positions and promotions), women appear to have fared as well as or better than men in the disciplines and type of institutions (RI) studied, and that they have had comparable access to many types of institutional resources (e.g., start-up packages, lab space, and research assistants). These findings are in contrast to the COSEPUP committee's general conclusions that "women who are interested in science and engineering careers are lost at every educational transition" and that "evaluation criteria contain arbitrary and subjective components that disadvantage women."

After providing a brief overview of the Status of Women in Academic Science and Engineering in 2004 and 2005 in Chapter 2, the report presents the results of the survey findings in the three areas: Academic Hiring (Chapter 3), Climate, Institutional Resources, Professional Activities, and Outcomes (Chapter 4), and Tenure and Promotion (Chapter 5). Chapter 6 provides an overall summary of key findings and recommendations, including questions for future research.

KEY FINDINGS

The surveys of academic departments and faculty have yielded interesting and sometimes surprising findings. **For the most part, men and women faculty in science, engineering, and mathematics have enjoyed comparable opportunities within the university, and gender does not appear to have been a factor in a number of important career transitions and outcomes.** The findings below provide key insights on gender differences in Academic Hiring (Chapter 3), Climate,

Institutional Resources, Professional Activities, and Outcomes (Chapter 4), and Tenure and Promotion (Chapter 5). Complete findings in each of these areas can be found at the end of the relevant chapter and are summarized in Chapter 6.

As a foundation for understanding the survey findings, it is important to remember that **although women represent an increasing share of science, mathematics, and engineering faculty, they continue to be underrepresented in many of those disciplines.** While the percent of women among faculty in scientific and engineering overall increased significantly from 1995 through 2003, the degree of representation varied substantially by discipline, and there remained disciplines where the percentage of women was significantly lower than the percentage of men. Table S-1 shows the percent of women faculty in selected scientific and engineering disciplines during this time period at the assistant, associate, and full professor levels.

In 2003, women comprised 20 percent of the full-time employed S&E workforce and had slowly gained ground compared to men in the full-time academic workforce; by 2003, they represented about 25 percent of academics. Women's representation in the academic workforce, of course, varied by discipline: in the health sciences, women were the majority of full-time, employed doctorates, while in engineering they were less than 10 percent. The greatest concentration of women among full-time academics was at medical schools; the lowest was at Research II institutions.

Academic Hiring (Chapter 3)

The findings on academic hiring suggest that many women fared well in the hiring process at Research I institutions, which contradicts some commonly held perceptions of research-intensive universities. If women applied for positions at Research I institutions, they had a better chance of being interviewed and receiving offers than male job candidates had. Many departments at Research I institutions, both public and private, have made an effort to increase the numbers and percentages of female faculty in science, engineering, and mathematics. Having women play a visible role in the hiring process, for example, has clearly made a difference. Unfortunately, women continue to be underrepresented in the applicant pool, relative to their representation among the pool of recent Ph.D.s. Institutions may not have effective recruitment plans, as departmental efforts targeted at women were not strong predictors in these surveys of an increased percentage of women applicants.

1. **Women accounted for about 17 percent of applications for both tenure-track and tenured positions in the departments surveyed. In each of the six disciplines, the percentage of applications from women for tenure-track positions was lower than the percentage of Ph.D.s awarded to women.** (Findings 3-1 and 3-3)

TABLE S-1 Representation of Women in Faculty Positions at Research I Institutions by Rank and Field (percent), 1995-2003

	Assistant Professor					Associate Professor					Full Professor				
	1995	1997	1999	2001	2003	1995	1997	1999	2001	2003	1995	1997	1999	2001	2003
Agriculture	17.8	18.6	19.6	18.1	27.2	12.7	12.5	10.7	17.6	13.9	4.9	5.2	6.1	6.6	8.0
Biology	35.6	38.2	36.0	37.0	38.8	26.0	24.3	26.3	30.2	31.2	14.0	14.7	15.8	18.0	20.8
Engineering	14.2	12.7	12.8	14.8	16.6	4.8	6.4	9.6	9.3	11.7	1.8	1.4	2.3	2.7	3.8
Health Sciences	69.1	66.9	64.0	64.7	66.5	65.6	65.1	64.9	64.5	59.1	35.1	38.9	45.3	48.0	59.0
Mathematics	18.7	22.0	26.5	25.2	26.6	10.4	14.4	14.9	15.8	16.3	7.6	5.9	9.9	10.0	9.7
Physics	25.1	25.6	24.6	25.4	24.1	9.5	13.4	14.8	16.7	19.5	4.3	4.6	5.9	6.8	7.6

SOURCE: National Science Foundation, Survey of Doctorate Recipients, 1995-2003. Tabulated by the National Research Council.

Table S-2 shows the percentage of women in the pool at each of several key transition points in academic careers: award of Ph.D., application for position, interview, and job offer. Although there was wide variation by field and department in the number and percentage of female applicants for faculty positions, the percentage of applications from women in each discipline was lower than the percentage of doctoral degrees awarded to women. This was particularly the case in chemistry and biology, the two disciplines in the study with the highest percentage of female Ph.D.s. The mean percentage of female applicants for tenure-track positions in chemistry was 18 percent, but women earned 32 percent of the Ph.D.s in chemistry from Research I institutions from 1999-2003. Biology (26 percent in the tenure-track pool and 45 percent in the doctoral pool) also showed a significant difference.

The fields with lower percentages of women in the Ph.D. pool had a higher propensity for those women to apply. Electrical engineering (11 percent in the tenure-track pool and 12 percent in the doctoral pool), mathematics, and physics, for example, had modest decreases in the applicant pool.

The percentage of applicant pools that included at least one woman was substantially higher than would be expected by chance. However, there were no female applicants (only men applied) for 32 (6 percent) of the available tenure-track positions and 16 (16.5 percent) of the tenured positions.

2. **The percentage of women who were interviewed for tenure-track or tenured positions was higher than the percentage of women who applied.** (Finding 3-10)

TABLE S-2 Transitions from Ph.D. to Tenure-Track Positions by Field at the Research I Institutions Surveyed (percent)

	Doctoral Pool	Pools for Tenure-Track Positions		
	Percent women Ph.D.s (1999-2003)	Mean percent of applicants who are women	Mean percent of applicants invited to interview who are women	Mean percent of first offers that go to women
Biology	45	26	28	34
Chemistry	32	18	25	29
Civil Engineering	18	16	30	32
Electrical Engineering	12	11	19	32
Mathematics	25	20	28	32
Physics	14	12	19	20

SOURCE: Survey of departments carried out by the Committee on Gender Differences in Careers of Science, Engineering, and Mathematics Faculty; Ph.D. data is from NSF, WebCASPAR.

For each of the six disciplines in this study the mean percentage of females interviewed for tenure-track and tenured positions exceeded the mean percentage of female applicants. For example, the female applicant pool for tenure-track positions in electrical engineering was 11 percent, and the corresponding interview pool was 19 percent.

3.   The percentage of women who received the first job offer was higher than the percentage who were invited to interview. (Finding 3-13)

For all disciplines the percentage of tenure-track women who received the first job offer was greater than the percentage in the interview pool. For example, women were 19 percent of the interview pool for tenure-track electrical engineering positions and received 32 percent of the first offers. This finding was also true for tenured positions with the notable exception of biology, where the interview pool was 33 percent female and women received 22 percent of the first job offers.

4.   Most institutional and departmental strategies for increasing the percentage of women in the applicant pool were not effective as they were not strong predictors of the percentage of women applying. The percentage of women on the search committee and whether a woman chaired the search, however, did have a significant effect on recruiting women. (Findings 3-7 and 3-8)

Departments have not generally been aggressive in using special strategies to increase the gender diversity of the applicant pool. Most of the policy steps proposed to increase the percentage of women in the applicant pool (such as targeted advertising, recruiting at conferences, and contacting colleagues at other institutions) were done in isolation, with almost two-thirds of the departments in our sample reporting that they took either no steps or only one step to increase the gender diversity of the applicant pool.

It appears that women were more likely to apply for a position if a woman chaired the search committee. The percentage of females on the search committee and whether a woman chaired the committee were both significantly and positively associated with the proportion of women in the applicant pool.

Professional Activities, Climate, Institutional Resources, and Outcomes (Chapter 4)

The survey findings with regard to climate and resources demonstrate two critical points. First, discipline matters, as indicated by the difference in the amount of grant funding held by men and women faculty in biology, but not in other disciplines. Second, institutions have been doing well in addressing most of the aspects of climate that they can control, such as start-up packages and reduced teaching loads. Where the challenge may remain is in the climate at the departmental level. Interaction and collegial engagement with one's colleagues is an important part of scientific discovery and collaboration, and here women faculty were not as connected.

5.   **Male and female faculty appeared to have similar access to many kinds of institutional resources, although there were some resources for which male faculty seemed to have an advantage.** (Findings 4-1 through 4-5)

Survey data revealed a great deal of similarity between the professional lives of male and female faculty. In general, men and women spent similar proportions of their time on teaching, research, and service; male faculty spent 41.4 percent of their time on teaching, while female faculty spent 42.6 percent. Male and female faculty members reported comparable access to most institutional resources, including start-up packages, initial reduced teaching loads, travel funds, summer salary, and supervision of similar numbers of research assistants and postdocs.

Men appeared to have greater access to equipment needed for research and to clerical support. At first glance, men seemed to have more lab space than women, but this difference disappeared once other factors such as discipline and faculty rank were accounted for.

6.   **Female faculty reported that they were less likely to engage in conversation with their colleagues on a wide range of professional topics.** (Findings 4-6, 4-7, and 4-8)

There were no differences between male and female faculty on two of our measures of inclusion: chairing committees (39 percent for men and 34 percent for women) and being part of a research team (62 percent for men and 65 percent for women). And although women reported that they were more likely to have mentors than men (57 percent for tenure-track female faculty compared to 49 percent for men), they were less likely to engage in conversation with their colleagues on a wide range of professional topics, including research, salary, and benefits (and, to some extent, interaction with other faculty members and departmental climate). This distance may prevent women from accessing important information and may make them feel less included and more marginalized in their professional lives. The male and female faculty surveyed did not differ in their reports of discussions with colleagues on teaching, funding, interaction with administration, and personal life.

7.   **There is little evidence across the six disciplines that men and women have exhibited different outcomes on most key measures (includ-**

ing publications, grant funding, nominations for international and national honors and awards, salary, and offers of positions in other institutions). The exception is publications, where men had published more than women in five of the six disciplines. On all measures, there were significant differences among disciplines. (Findings 4-9 through 4-14)

Overall, male faculty published marginally more refereed articles and papers in the past 3 years than female faculty, except in electrical engineering, where the reverse was true. Men published significantly more papers than women in chemistry (men: 15.8; women: 9.4) and mathematics (men: 12.4; women: 10.4). In electrical engineering, women published marginally more papers than men (men: 5.8; women: 7.5). The differences in the number of publications between men and women were not significant in biology, civil engineering, and physics.

There were no significant gender differences in the probability that male or female faculty would have grant funding, i.e., be a principal investigator or co-principal investigator on a grant proposal. Male faculty had significantly more research funding than female faculty in biology; the differences were not significant in the other disciplines.

Female assistant professors who had a mentor had a higher probability of receiving grants than those who did not have a mentor. In chemistry, female assistant professors with mentors had a 95 percent probability of having grant funding compared to 77 percent for those women without mentors. Over all six fields surveyed female assistant professors with no mentors had a 68 percent probability of having grant funding compared to 93 percent of women with mentors. This contrasts with the pattern for male assistant professors; those with no mentor had an 86 percent probability of having grant funding compared to 83 percent for those with mentors.

Male and female faculty were equally likely to be nominated for international and national honors and awards, although the results varied significantly by discipline. Gender was a significant determinant of salary among full professors; male full professors made, on average, about 8 percent more than females, once we controlled for discipline. At the associate and assistant professor ranks, the differences in salaries of men and women faculty disappeared.

Tenure and Promotion (Chapter 5)

The findings related to tenure and promotion indicate the importance of addressing the retention of women faculty in the early stages of their academy careers; not as many were considered for tenure as would be expected, based on the number of women assistant professors. Retention was particularly problematic given the increased duration of time in rank for all faculty. Both male and female faculty utilized stopping-the-tenure-clock policies—spending a longer time in the uncertainty of securing tenure—but women used these policies more. Women faculty who did come up for tenure were as successful or more successful than men, so one of the most important challenges may be in increasing the pool of women faculty who make it to that point.

8. **In every field, women were underrepresented among candidates for tenure relative to the number of female assistant professors. Most strikingly, women were most likely to be underrepresented in the fields in which they accounted for the largest share of the faculty— biology and chemistry.** (Finding 5-1)

In biology and chemistry, the differences were statistically significant. In biology, 27 percent of the faculty considered for tenure were women, while women represented 36 percent of the assistant professor pool. In chemistry those numbers were 15 percent and 22 percent, respectively. This difference may suggest that female assistant professors were more likely than men to leave before being considered for tenure. It might also reflect the increased hiring of female assistant professors in recent years (compared with hiring 6 to 8 years ago).

9. **Women were more likely than men to receive tenure when they came up for tenure review.** (Findings 5-2, 5-3, and 5-4)

In each of the six fields surveyed, women were tenured at the same or a higher rate than men (an overall average of 92 percent for women and 87 percent for men). It appears that women were more likely to be promoted when there was a smaller percentage of females among the tenure-track faculty. Discipline, stop-the-tenure-clock policies, and departmental size were not associated with the probability of a positive tenure decision for either male or female faculty members who were considered for tenure. Both male and female assistant professors were significantly more likely to receive tenure at public institutions (92 percent) than at private institutions (85 percent).

10. **No significant gender disparity existed at the stage of promotion to full professor.** (Findings 5-6 and 5-7)

For the six disciplines surveyed, 90 percent of the men and 88 percent of the women proposed for full professorship were promoted—a difference that was not statistically significant, after accounting for other potentially important factors such as disciplinary differences, departmental size, and use of stopping-the-tenure-clock policies. Women were proposed for promotion to full professor at approximately the same rates as they were represented among associate professors.

11. Women spent significantly longer time in rank as assistant professors than did men. (Findings 5-8 and 5-9)

Although time in rank as an assistant professor has increased over time for both men and women, women showed significantly longer durations than men. It is difficult to determine whether these apparent differences may be explained, at least in part, by individual and departmental characteristics such as length of postdoctoral experience and stopping-the-tenure-clock for family leave. Both male and female faculty spent more time in the assistant professor ranks at institutions of higher prestige.

12. Male and female faculty who stopped the tenure clock spent significantly more time as assistant professors than those who did not (an average of 74 months compared to 57 months). They had a lower chance of promotion to associate professor (about 80 percent) at any time (given that they had not been promoted until then) than those who did not stop the clock. Everything else being equal, however, stopping the tenure clock did not affect the probability of promotion and tenure; it just delayed it by about 1.5 years. It is unclear how that delay affected women faculty, who were more likely than men to avail themselves of this policy. (Finding 5-10)

Although the effect of stopping the tenure clock on the probability of promotion and tenure is similar for both male and female faculty, 19.7 percent of female assistant professors in the survey sample availed themselves of this policy compared to 7.4 percent of male assistant professors. At the associate professor level, 10.2 percent of female faculty compared to 6.4 percent of male faculty stopped the tenure clock.

RECOMMENDATIONS

The survey data suggest that positive changes have happened and continue to occur. At the same time, the data should not be mistakenly interpreted as indicating that male and female faculty in math, science, and engineering have reached full equality and representation, and we caution against premature complacency. Much work remains to be done to accomplish full representation of men and women in academic departments.

Many of the survey findings point out specific areas in which research institutions and professional societies can enhance the likelihood that more women will apply to faculty positions and persist in academia up to and beyond tenure and promotion. Changes in the faculty recruitment and search process, enhancement of mentoring programs, broader dissemination of tenure and stop-the-tenure-clock policies, and investigation of the subtle effects of climate on career decisions can all help. Increased data collection, of course, is also necessary. Specific recommendations for institutions and professional societies are delineated in Chapter 6.

QUESTIONS FOR FUTURE RESEARCH

This study raises many unanswered questions about the status of women in academia. As noted at the onset of this report, the surveys did not capture the experiences of Ph.D.s who have never applied for academic positions, nor of female faculty who have left at various points in their academic careers. We also recognize that there are important, nonacademic issues affecting men and women differentially that impact career choices at critical junctures. Fuller examination of these issues (for example, topics relating to family, children, home life, care of elderly parents) will shed greater light on career choices by women and men and should yield suggestions on the types of support needed to encourage retention of women in academic careers. Below are suggestions for future research:

A Deeper Understanding of Career Paths

1. Using longitudinal data, what are the academic career paths of women in different science and engineering disciplines from receipt of their Ph.D. to retirement?

2. Why are women underrepresented in the applicant pools and among those who are considered for tenure?

3. Why aren't more women in fields such as biology and chemistry applying to Research I tenure-track positions, as discussed in Finding 3-3?

4. Why do female faculty, compared to their male counterparts, appear to continue to experience some sense of isolation in more subtle and intangible areas?

5. What is the impact of stop-the-tenure-clock policies on faculty careers?

6. What are the causes for the attrition of women and men prior to tenure decisions, if indeed attrition does take place?

7. To what extent are women faculty rewarded beyond promotion to full professor?

8. What important, nonacademic issues affect men and women differentially that impact their career choices at critical junctures?

Expanding the Scope

9. How important are differences among fields?

10. What are the experiences of faculty at Research II institutions?

11. What are the experiences of part-time and nontenure track faculty?

1

Introduction

The 1999 report, *A Study on the Status of Women Faculty in Science at MIT*,[1] created a new level of awareness of the special challenges faced by female faculty in the sciences. Although not the first examination of the treatment of female faculty, this report marked an important historical moment, igniting interest in the difficulties experienced by many women, particularly those at the higher levels of academia. Since the release of the Massachusetts Institute of Technology report, many other institutions have studied equity issues regarding their faculty, and several have publicly pledged to use their resources to correct identified disparities. Although academic departments, institutions, professional societies, and others have paid more attention to the topic in the past 10 years, there has been concern that remedial actions have approached a plateau.

Unquestionably, women's participation in academic science and engineering (S&E) has increased over the past few decades. In the 10 years prior to the start of this study, the number of women receiving Ph.D.s in science and engineering increased from 31.7 percent (in 1996) to 37.7 percent (in 2005).[2] The percentage of women among doctoral scientists and engineers employed full-time, while still small, rose from 17 percent in 1995 to 22 percent in 2003.[3] However, women continued to be underrepresented among academic faculty relative to the number of women receiving S&E degrees. In 2003, women comprised between 18 and 45

---

[1] See Massachusetts Institute of Technology (1999).

[2] National Science Foundation (2006); Figure A2-1 and Table A2-1 in Appendix 2-1.

[3] National Science Foundation, Survey of Doctorate Recipients, 1995-2003; Figure A2-3 in Appendix 2-1.

15

percent of assistant professors in S&E and between 6 and 29 percent of associate and full professors.[4]

The evidence for disparities in the treatment of women and men is mixed. In some cases (e.g., with regard to salaries), there are strong quantitative data. In other cases (e.g., marginalization), the evidence is more anecdotal. Still in other instances, the evidence is scant or missing. Assessing whether search committee members are biased in their evaluations of male and female candidates could be—and has been—done in essentially a laboratory-like setting, but there are no publicly available national data upon which to draw.

WHY DISPARITIES MATTER

Interest in studying the disparities between the careers of male and female faculty is widespread. Government agencies, legislators, and organizations, including many professional societies, have a vested interest in promoting science and engineering education and careers and encouraging a diverse set of students and graduates to enter and remain in S&E. Administrators in the academic community need benchmarks to help set the context in which universities conduct their own self-examinations—as many already do. S&E students considering academia among their career options are seeking better information about career prospects and challenges.

Why is an assessment needed now? Three reasons support this.[5] First, the nature of the academic profession is changing in several important ways, including the composition of the profession, reward structure, and professional activities. Due in part to the diminishing financial resources and increasing costs faced by higher education institutions, hiring into tenure-track positions has slowed, while the number of part-time, temporary, and off-track positions has increased. Such changes may affect female academics differently than male academics.

Second, substantial efforts to increase women's participation as faculty in higher education have been underway for three decades. These include programs and policies of the federal government, professional societies, and their universities and individual academic departments. At the federal level, one example is the National Science Foundation's (NSF's) ADVANCE program. Scientific and professional societies focused on women generally or in specific disciplines have collected relevant data and undertaken programs to support women in the profession (e.g., the Association for Women in Science [AWIS], the Society of Women Engineers [SWE], the Committee on the Status of Women in Physics [CSWP], and the Caucus for Women in Statistics). Higher education institutions have conducted

---

[4] See Tables 2-1 and 2-2.

[5] See also the four reasons suggested by NAS, NAE, and IOM (2007): global competitiveness, law, economics, and ethics.

gender equity studies and developed work-life policies for faculty and staff.[6] An assessment of changes in faculty composition as well as policies and outcomes related to faculty careers is one step in evaluating these efforts.

Finally, where gender disparities exist and women are underrepresented among S&E faculty, negative consequences result that require policy solutions. Substantial resources go into producing a Ph.D. in S&E.[7] The untapped potential of fully trained and credentialed women, as well as the women who are interested in S&E but choose not to pursue degrees because of obstacles, real or perceived, represents an important economic loss—one a competitive United States cannot afford. As Senator Ron Wyden (2003) stated:

> A report from the Hart-Rudman Commission on National Security to 2025 warned that America's failure to invest in science and to reform math and science education was the second biggest threat to our national security, greater than that from any conceivable conventional war. America will not remain the power it is in the world today, nor will our people be as healthy, as educated, or as prosperous as they should be, if we do not lead the world in scientific research and engineering development. To make our country better, to improve our national security and quality of life, we need to encourage people to go into these disciplines. Women represent a largely untapped resource in achieving this vital goal.

Similarly, Neal Lane, former Assistant to the President for Science and Technology, remarked to the Summit on Women in Engineering (1999) that "we simply need people with the best minds and skills, and many of those are women." This view was echoed by leaders of nine top research universities in a meeting at MIT in 2001 to discuss women faculty in science and engineering. A joint statement issued by the participants noted, "Institutions of higher education have an obligation, both for themselves and for the nation, to fully develop and utilize all creative talent available. We recognize that barriers still exist to the full participation of women in science and engineering" (Campbell, 2001b).

A more inclusive workforce may be more innovative and productive than one which is less so. As Arden L. Bement, Jr., Director of the National Science Foundation, said in 2005:

> Year by year, the economic imperative grows for broadening, empowering, and sharpening the skills of the entire U.S. workforce—just to remain competitive in the global community. This fresh talent is our most potent mechanism for technology transfer to our systems of innovation. Fortunately, we have a fount of untapped talent in our women, underrepresented minorities and persons with

---

[6] For a list of gender equity studies conducted by Research I institutions, see the CWSEM Web site at http://www.nas.edu/cwsem.

[7] The average annual support for a doctoral student is $50,000 according to a new study (NAS, NAE, and IOM, 2007). The average doctoral student takes 7 years to complete a Ph.D., suggesting support for a single student could be $350,000.

disabilities. Our need to broaden participation and increase opportunity is critical, for both the science and education communities and the nation.[8]

"Having scientists and engineers with diverse backgrounds, interests, and cultures assures better scientific and technological results and the best use of those results." (Lane, 1999). If, for example, women approach the process of S&E teaching or research differently or generate different, important outcomes (findings, publications, patents, etc.), then their relative exclusion somewhat diminishes the potential of academia (Xie and Shauman, 2003:footnote 2). A comparison of data from the National Survey of Student Engagement (NSSE) and the Faculty Survey of Student Engagement (FSSE) indicates that when faculty emphasized effective educational practices, students tended to engage more in those practices. Interestingly, the FSSE found women were more likely than men to value and use effective educational practices (Kuh et al., 2004).

"Academic institutions play a pivotal role in preparing the science and engineering work force, and their faculty and leaders serve as intellectual, personal, and organizational role models that shape the expectations of future scientists and engineers," said Alice Hogan, NSF's former ADVANCE Program Manager. "Ensuring that the climate, the policies and the practices at these institutions encourage and support the full participation of women in all aspects of academic life, including leadership and governance, is critical to attracting students to science and engineering careers" (Harms, 2001).

Women are students before they enter the workforce. Female faculty, by acting as role models, produce the next generation of scholars and are associated with greater production of female S&E students. According to Trower and Chait (2002:34), the "most accurate predictor of subsequent success for female undergraduates is the percentage of women among faculty members at their college."

Finally, there are legal prescriptions prohibiting discrimination and questioning the propriety of disparities (see NAS, NAE, and IOM, 2007 for a review of antidiscrimination laws). The Equal Pay Act of 1963, Title VII of the Civil Rights Act of 1964, and Title IX of the Education Amendments of 1972 all focus on prohibiting sex discrimination. Title IX is a particularly relevant piece of legislation, prohibiting discrimination on the basis of sex in federally assisted education programs or activities. Most frequently invoked to promote equal access to athletic programs, Title IX also covers employment, and a 2004 Government Accountability Office (GAO) report suggested efforts to enforce compliance with Title IX should be applied more broadly to educational institutions. The Science and Engineering Equal Opportunities Act of 1980 declares "it is the policy of the United States that men and women have equal opportunity in education, training and employment in scientific and technical fields." As Lane (1999) noted, "Careers

---

[8] Arden L. Bement, Jr., "Remarks, Setting the Agenda for 21st Century Science," at the meeting of the Council of Scientific Society Presidents, December 5, 2005. Available at http://www.nsf.gov/news/speeches/bement/05/alb051205_societypres.jsp.

in science and engineering are immensely rewarding, and all Americans should have the opportunity to participate—it's what America is all about."

THE COMMITTEE'S CHARGE

The concern that inequities still exist, as well as the need for empirical evidence to conduct a search for disparities, prompted this study. In 2002, Senator Ron Wyden (D-Oregon), of the Subcommittee on Science, Technology and Space of the U.S. Senate Committee on Commerce, Science and Transportation convened three hearings on the subject of women studying and working in science, mathematics, engineering, and technology.[9] Soon after, Congress directed the NSF to contract with the National Academies for a study assessing gender differences in the careers of science and engineering faculty, based on both existing and new data.[10]

To meet this charge, the National Academies appointed an ad hoc study committee—the Committee on Gender Differences in Careers of Science, Engineering, and Mathematics Faculty—to examine this issue under the auspices of the Committee on Women in Science and Engineering (CWSE) and the Committee on National Statistics (CNSTAT). (Appendix 1-1 identifies the members of the study committee and describes their areas of expertise.) The committee was guided by the following statement of task:

> An ad hoc committee will conduct a study to assess gender differences in the careers of science, engineering, and mathematics (SEM) faculty, focusing on four-year institutions of higher education that award bachelor's and graduate degrees. The study will build on the Academy's previous work and examine issues such as faculty hiring, promotion, tenure, and allocation of institutional resources including (but not limited to) laboratory space.

APPROACH AND SCOPE

Approach

The committee interpreted its charge to include three goals: (1) to update earlier analyses with newer information, (2) to provide a more thorough understanding of the scope of potential gender differences in S&E faculty, and (3) to recommend methods for further informing or clarifying assumptions about gender and academic careers. Establishing causes for any observed differences, while an

---

[9] See Statement of Senator Ron Wyden, Hearing on Title IX and Science, U.S. Senate Committee on Commerce, Science and Transportation, October 3, 2002.

[10] In addition to this activity, the Government Accountability Office was asked to complete a study on Title IX (GAO, 2004), and the RAND Corporation conducted a study on gender differences in federal funding (Hosek et al., 2005).

important task, was considered to be beyond the scope of the charge. For purposes of this report, science and engineering are defined as the physical sciences (including astronomy, chemistry, and physics); earth, atmospheric, and ocean sciences; mathematics and computer science; biological and agricultural sciences; and engineering (in all its forms).[11]

The committee understood the charge as focusing primarily on major research universities—known as the Research I (RI) or research-intensive institutions—for several reasons.[12] First, the committee believed gender disparities, if present, are more likely to occur in these institutions. Second, findings for research universities are likely to serve as a good starting point for the consideration of gender disparities in other sectors of higher education. Finally, and most important, as is discussed more fully below, research universities play especially important roles in training doctoral students and future scholars and faculty.

Recognizing at the outset the need for new data, the committee conducted two national surveys in 2004 and 2005 of faculty and academic departments in six science and engineering disciplines: biology, chemistry, civil engineering, electrical engineering, mathematics, and physics. The first survey of almost 500 departments focused on hiring, tenure, and promotion processes, while the second survey gathered career-related information from more than 1,800 faculty. Together the surveys addressed departmental characteristics, hiring, tenure, promotion, faculty demographics, employment experiences, and types of institutional support received. In addition to results from the surveys, the committee heard expert testimony and examined data from federal agencies and professional societies, individual university studies (e.g., gender equity, salary, or "climate" studies), and academic articles. The survey is discussed in greater detail later in this chapter and in Appendix 1-4.

---

[11] The term "sciences and engineering" is often defined as the academic disciplines of physical sciences (including astronomy, chemistry, and physics); earth, atmospheric, and ocean sciences; mathematical and computer sciences; biological and agricultural sciences; and engineering (in all its forms). Additionally, psychology and the social sciences (including economics, political science, and sociology) may also be treated as science fields. Non-S&E fields are defined to include the various arts and humanities. The natural sciences and engineering are defined in this study as agricultural sciences, biological sciences, health sciences, engineering, computer and information sciences, mathematics, and physical sciences. Further gradations can be seen in the Survey of Earned Doctorates list of fields of study. Our definition includes Ph.D. fields coded as between 005 and 599, inclusive. Refer to the questionnaire, an example of which is found at http://www.nsf.gov/statistics/nsf06308/pdf/nsf06308.pdf.

[12] Research I institutions are defined as institutions which offer, beyond baccalaureate programs, doctoral programs which award 50 or more doctoral degrees annually. In addition these institutions receive a substantial amount ($40 million or more) of federal support. Note that this definition is based on the 1994 classification devised by The Carnegie Foundation for the Advancement of Teaching. The classification scheme was redone in 2000 and 2005. See "Carnegie Classifications" at http://www.carnegiefoundation.org/classifications/ for further details.

There is no question that academic careers vary significantly for both men and women, depending on the type of academic institution and the academic position, so the findings from these surveys may or may not be relevant to other academic appointments or institutions. While by no means exhausting the topic, the purpose of this report is to advance the state of knowledge on specific aspects of gender in academic science and engineering, while at the same time recognizing the study's limitations.

There are many factors that play a significant role in women's careers in academia that are outside the charge and therefore were excluded in the committee's deliberations. These include, for example:

- Constraints of dual careers, particularly in geographic mobility;
- Access to quality child care;
- Impact of stopping-the-tenure-clock policies;
- Preference for part-time academic positions;
- Perceptions of isolation and lack of collegiality;
- Expectations regarding professional recognition and career satisfaction;
- Attrition along the academic career pathway;
- Disciplinary differences that either foster or impede these factors; and
- Other quality-of-life issues.

In particular, the report does not explore the impact of children and family life. While these and similar factors are beyond the scope of this study, they are significant in impacting women's faculty career choices.

Also, incremental changes in the percentages of women with doctoral degrees and in postdoctoral positions do not by themselves result in commensurate changes in the numbers of women faculty in universities, especially at senior levels. Much more needs to be known about the careers of women scientists after and even during graduate school, as well as the many career paths they may follow that may lead them away from academia. This study focuses primarily on key transition points in academic careers that research-intensive institutions can control and influence. Substantial additional research is needed to create a more complete picture of women's career paths (see suggestions in Chapter 6).

The study reassesses and extends, with newly collected data, results of prior examinations of gender differences in academia to establish the contemporary veracity of those conclusions and to document trends over time. The study moves beyond earlier analyses by focusing more directly on the role of three sets of factors thought to produce gender differences in academic careers: (1) institutional practices and procedures, including the hiring and tenure processes; (2) individual characteristics, such as the role of marriage and family in the academic career paths of men and women; and (3) the overarching, changing nature of the academic profession. Focusing on these factors, the committee reformulated the charge into a series of guiding research questions about academic hiring, institutional resources and climate, and tenure and promotion.

Academic Hiring (Chapter 3)

- Is gender associated with the probability of individuals applying for S&E positions in Research I institutions?
- Given that an individual applies for a position, does a woman have the same probability of being interviewed as a man?
- Given that an individual is interviewed for a position, does a woman have the same probability of being offered a position as a man?

Institutional Resources, Professional Activities, and Climate (Chapter 4)

- Do male and female faculty engage in similar professional activities?
- Do male and female faculty receive similar institutional resources?
- Are male and female faculty similarly productive in terms of research?
- Is the departmental/institutional climate the same for male and female faculty?
- Do male and female faculty have similar rates of retention and degrees of job satisfaction?

Tenure and Promotion (Chapter 5)

- Are similar male and female faculty equally likely to receive tenure?
- Are similar male and female faculty equally likely to receive a promotion?
- Do men and women spend similar amounts of time at lower and intermediate ranks?

To answer these questions, the committee relied on multiple sources of information, but especially on information collected through two national surveys of individual faculty and academic departments, described in detail later in this chapter. Chapters 3, 4, and 5 present the results of the statistical analyses of the data collected in the surveys during the course of this study. In a number of cases, findings from the current surveys differ from some of the positions put forth in the literature, as summarized in Chapter 2. Recommendations offered in Chapter 6 are based directly on the committee's analysis of the survey data.

Scope

This study is necessarily limited. Academia in the United States is both broad and varied, and the factors affecting the career tracks of female Ph.D.s in science and engineering are diverse and complex. This report focuses on a small but vital segment of higher education, a specific population of faculty members, and factors affecting academic careers largely controlled by institutions. It does not cover all of higher education, all faculty members, or all factors affecting career tracks or decisions. Put succinctly, the report examines key institutional transitions and experiences of male and female, full-time, assistant, associate, and full professors in the natural sciences and engineering at Research I institutions.

What Career Factors Are Examined

As is readily apparent to anyone who has studied, considered, or experienced an academic career, many vital transition points and factors affect career choices and decisions. These encompass influences from as early as high school or middle school to decisions and opportunities until (and beyond) retirement. They include decisions or opportunities to pursue academic careers, work in industry or government, or take oneself out of the job market. They cover, of course, formal institutional actions, such as those described here, as well as unofficial and unstated actions difficult to measure. And they include a myriad of personal characteristics, family circumstances, social pressures, opportunities, and experiences of female faculty members and those who might have become faculty. Many of the "whys" of the findings included here are buried in factors that the committee was unable to explore.

We do not know, for example, what happens to the significant percentage of female Ph.D.s in science and engineering who do not apply for regular, faculty positions at Research I institutions. Do they pursue faculty jobs at other universities or colleges? Become clinical, adjunct, or research faculty members or other research personnel? Get postdocs? Take positions in industry or government? Opt out of the workforce altogether? Some factors to consider are:

Presence of role models and mentors
Finances
Parental influence
Family circumstances
Professional networks
Job market
Geographical restrictions

In the same vein, we do not know what happens to women faculty members who are hired and subsequently leave the university. The entire range of options available to new Ph.D.s is available to them, in addition to many institutional factors, such as:

Salary level
Likelihood of promotion

Denial of tenure
Institutional funding
Personal affinity for teaching or research
Family circumstances
Institutional climate
Productivity
Social factors

For those who remain in regular faculty positions, the report does include important and new information on their individual characteristics, family circumstances, professional activities, and outcomes, as well as institutional resources and climate. But even for this group, there are many factors affecting individual choices and institutional climate that we were unable to measure.

At the senior end of the academic career track, we know little about female full professors and what gender differences might exist at this stage of one's career. This report does not include descriptions of special institutional programs or recognitions such as:

Salary adjustments
Research support
Named chairs or professorships
Leadership positions

Who and What Are Included

In addition to focusing on select factors affecting academic careers, the study has limited its scope to particular types of institutions, individuals, and disciplines. First, the focus of this study is primarily current, rather than historical or predictive. It is beyond the scope of the charge and the resources of the committee overseeing this report to estimate future trends for female faculty.

Second, there are thousands of higher education institutions in the United States. This study does not address any pipeline issues regarding educational preparation and training prior to application for a tenure-track position. As stated above, the study focuses primarily on doctoral-granting institutions, specifically the 89 Research I institutions (also know as research-intensive institutions) defined by The Carnegie Foundation for the Advancement of Teaching in 1994 and listed in Appendix 1-2. These institutions were picked because of their prestige, the role they play in training future generations of scholars, their contribution to scholarship, and the amount of research they undertake.[13] The data gathered

---

[13] The National Science Foundation (2002:2-3) notes: "Research universities enroll only 19 percent of the students in higher education, but they play the largest role in S&E degree production. They produce most of the engineering degrees and a large proportion of natural and social science degrees

about research universities will also likely serve as a useful starting point for the examination of other types of higher education institutions.

Third, this study will focus primarily on full-time, regularly appointed, professorial faculty. Due to the committee's interest in what has traditionally been the typical academic career path within Research I institutions, the target population is limited to assistant, associate, and full professors. By and large, these are the faculty who are tenure eligible, who both teach and conduct research, who supervise most of the graduate students who will be the next generation of scholars, and who are most likely to receive the widest range of institutional support. Instructors, lecturers, postdocs, adjunct faculty, clinical faculty, and research faculty are not included. While these faculty are important, they have very different career paths and warrant separate study.

Fourth, although data are provided for many natural science and engineering disciplines in assessing historical gender differences in academia, the new data collected for this report by the two surveys of department chairs and faculty focus on six fields: the biological sciences, chemistry, civil engineering, electrical engineering, mathematics, and physics.[14] The purpose of the primary data collection on a subset of fields was to allow for an examination of the career paths for men and women facing similar expectations and constraints. Although the findings may identify male/female differences prevalent throughout science and

---

at both the graduate and undergraduate levels. In 1998, the nation's 127 research universities awarded more than 42 percent of all S&E bachelor's degrees and 52 percent of all S&E master's degrees." For example, of the 8,350 Ph.D.s granted in the life sciences in 2002, 2,608 Ph.D.s (31 percent) were granted by just 20 Research I institutions (Hoffer et al., 2003). These institutions "are also the most conducive organizational contexts for a prestigious research career" (NRC, 2001a:124). On federal academic S&E support, see Richard J. Bennof, *Federal Science and Engineering Obligations to Academic and Nonprofit Institutions Reached Record Highs in FY 2002*, NSF InfoBrief, June 2004, (NSF 04-324).

[14] The four science fields were chosen, partly because they represent the "standard" or well-known science fields. In addition, professional associations in the areas of chemistry, mathematics, and physics collect data on their fields. Readers should note that "biological sciences" is a broad term, and may include agricultural or health sciences. Likewise, mathematics data sometimes include data for statistics or computer science. Finally, physics data may include astronomy.

Civil engineering was chosen as a middle ground among the various engineering fields. According to Gibbons (2004), during the 2002-2003 academic year, more than 8,000 students received civil engineering baccalaureate degrees—the fourth largest amount—and women received 23.4 percent of those degrees. This lies between a high for environmental engineering (42.1 percent of degrees went to women) and a low of 11.7 percent for engineering technology. About 3,600 students received master's degrees—the fifth largest amount—and women received 25.2 percent of them, between 42.2 percent for environmental engineering and 9.0 percent for petroleum. The third largest amount— 631 doctoral degrees were awarded and women received 18.4 percent of them, between 33.3 percent for engineering management and zero percent in mining and in architectural engineering. Finally, for faculty, civil engineering had the third highest number of faculty members: 3,320, and 10.9 percent of tenured/tenure-track teaching faculty were women. Fields with the lowest percentage of women were aerospace, petroleum, and mining (all at 5.0 percent); while the highest were biomedical (16.6 percent), industrial/manufacturing (15.4 percent), and environmental (14.7 percent).

engineering faculties, the reader is cautioned about generalizing from the findings. Not only may they not apply to all fields of science and engineering, but also it may be inappropriate to generalize from findings in physics and chemistry, for example, to all physical sciences or from civil and electrical engineering to all engineering fields.

Differences and Commonalities with Other National Academies' Reports

The committee has benefited greatly from three other National Academies' reports on women in academic science and engineering. In 2001 NRC published *From Scarcity to Visibility: Gender Differences in the Careers of Doctoral Scientists and Engineers,*[15] a statistical analysis of the career progression of matched cohorts of men and women Ph.D.s from 1973 to 1995, using data from the NSF Survey of Earned Doctorates and Survey of Doctoral Recipients. The 2001 report had a much broader scope than this one; it covered employment outside academia; all science and engineering disciplines including the social sciences; and (within academia) all types of higher education institutions and faculty positions. It relied on longitudinal data on the same individuals collected over time, rather than a snapshot of faculty and departments at a single point in time. While it is not possible to draw direct comparisons between the data in the two reports, some of the 2001 findings on women's participation in academia provide a useful backdrop:

- Men hold a 14 percent advantage in tenure-track positions.
- Women are underrepresented in senior faculty positions at Research I institutions.
- At any professional age, men are more likely than women to hold tenure.
- Women are less likely to be full professors than are their male counterparts.

The 2005 NRC report, *To Recruit and Advance: Women Students and Faculty in U.S. Science and Engineering,*[16] identifies the strategies that higher education institutions have employed to achieve gender inclusiveness, based on case studies of four successful universities. Concluding that women face "challenges that may lead to their attrition at key junctures in higher education" and that "female faculty appear to advance along the academic career pathway more slowly than males," the 2005 report identifies successful strategies for recruitment and retention of women undergraduate and graduate students, recruitment and advance-

---

[15] National Research Council, 2001, *From Scarcity to Visibility: Gender Differences in the Careers of Doctoral Scientists and Engineers.* Washington, DC: National Academy Press.

[16] National Research Council, 2005, *To Recruit and Advance: Women Students and Faculty in U.S. Science and Engineering,* Washington, DC: National Academies Press.

ment of women faculty, and advancement of women faculty into administrative positions.

A third report, *Beyond Bias and Barriers: Fulfilling the Potential of Women in Academic Science and Engineering*, was released in 2006.[17] Appointed under the aegis of the Committee on Science, Engineering, and Public Policy (COSEPUP), this study committee was charged to "review and assess the research on sex and gender issues in science and engineering, including innate differences in cognition, implicit bias, and faculty diversity" and to "provide recommendations to guide faculty, deans, department chairs, other university leaders, funding organizations, and government agencies in the best ways to maximize the potential of women science and engineering researchers."

Beyond Bias and Barriers examines the results of recent research on gender differences in learning and performance—particularly cognitive, biological, and sociocultural differences that address the educational pathways to becoming faculty. It lists 11 common beliefs about women in science and engineering and presents evidence refuting them. Based primarily on existing data and the committee's expertise, it identifies barriers that women face in academia and calls for action by university leaders, professional societies, federal agencies, and Congress to "transform institutional structures and procedures to eliminate gender bias."

The COSEPUP report is significantly broader in scope than this report. It covers faculty from all fields of sciences and engineering (including the social sciences) and encompasses the full range of academic institutions. It addresses the overall mobility of women in academia, as well as the specific concerns of minority women. And based on an assessment of the underlying causes of gender discrepancies in academia, it provides broad policy recommendations for changes at higher education institutions.

In contrast, and following COSEPUP's recommendation for new and accurate information, this report examines the experiences of a specific set of faculty and departments in six disciplines in a particular type of institution (Research I), based primarily on data collected in 2004 and 2005. Rather than an overview of career paths, our examination is limited to a snapshot of key transition points in academic careers that are under the control of the institutions (hiring, institutional climate and resources, tenure, and promotion). It highlights many striking differences among the disciplines that make generalizations across science and engineering difficult. The findings and recommendations here are a direct result of the data from our two surveys, which were not available to the COSEPUP committee.

Given the differences in scope and approach, it is not surprising that some of the findings of the two reports differ. While both committees found that women are underrepresented in academic science and engineering, the survey findings presented here indicate that at many critical transition points in their academic

---

[17] National Academies, 2007, *Beyond Bias and Barriers: Fulfilling the Potential of Women in Academic Science and Engineering*. Washington, DC: National Academies Press.

careers (e.g., hiring for tenure-track and tenured positions and promotions), women appear to have fared as well as or better than men in the disciplines and type of institutions (Research I) studied. The survey data show that female and male faculty have had comparable access to many types of institutional resources (e.g., start-up packages, laboratory space, and research assistants), in contrast to the COSEPUP committee's general findings that "women who are interested in science and engineering careers are lost at every educational transition"[18] and that "evaluation criteria contain arbitrary and subjective components that disadvantage women."[19]

Like the COSEPUP committee, however, this committee found evidence of the overall loss of women's participation in academia, even though many of the actual transition points under the control of institutions (like interviewing, hiring, and promoting) do not show evidence of a loss. The loss is most apparent in the smaller fraction of women who apply for faculty positions and in the attrition of female assistant professors before tenure consideration. The former is especially apparent in the fields of chemistry and biology, where the number of female applicants for faculty positions in Research I institutions is much lower than the number of women doctorates in the pool. Unfortunately, our surveys do not shed light on why women fail to apply for faculty positions or why (or if) they leave academia between these critical transition points. Similarly, the reports agree that there are gender differences in time in rank, but we do not have any causal evidence as to why this is so.

The findings in both reports underscore the fact that our work is not done. Further research is needed, along with continued efforts to increase the number of women faculty in many disciplines and at key points in academic careers.

Sources of Information

The primary source of information for this report consists of two new surveys designed and conducted especially for this project by the American Institute of Physics during 2004 and 2005. The surveys were undertaken to fill in some of the current gaps in knowledge regarding faculty outcomes and institutional practices, which could not otherwise be addressed by existing data sets. One survey focused on departments; the other examined faculty.

The departmental survey was a census of biology, chemistry, civil engineering, electrical engineering, mathematics, and physics departments at Research I institutions ($N = 492$). It gathered information on departmental characteristics, hiring practices and outcomes, and tenure and promotion processes and yielded an overall response rate of 85 percent. Data on attrition were not collected.

---

[18] Ibid, p. 2.
[19] Ibid, p. 3.

In contrast, the faculty survey was a stratified, random sample of approximately 1,800 faculty from the same departments. The faculty survey included information on demographic characteristics, employment experiences, and types of institutional support received and yielded a response rate of 73 percent. Comparable, cross-institution information on hiring and resource allocation is notoriously difficult to find—although some universities collect such information—and thus the survey data collected for this project is quite instructive. Because of funding limitations and concern that longer surveys would have lower response rates, the surveys neither included questions about degree of job satisfaction nor collected information on attrition of faculty over the preceding several years. Hopefully, others will collect some of the information that could not be gathered in the course of this study. Details on the implementation of the surveys, including the actual questionnaires and response rates, can be found in Appendix 1-4 and Appendix 1-5.

To gain a better understanding of the overall representation of women in academic science and engineering and how that has changed over time, the committee examined data from two large, national studies: the Survey of Doctoral Recipients (SDR), conducted biennially by the NSF, and the National Survey of Postsecondary Faculty, conducted every five years by the National Center for Education Statistics of the Department of Education. Data from professional and disciplinary societies were also examined.

To determine the state of current knowledge on women's academic career paths, the committee reviewed studies conducted by individual universities as well as publications by individual researchers. It also heard expert testimony from several interested stakeholders at its first committee meeting (see Appendix 1-3).

Drawing from these multiple sources, Chapter 2 provides a brief overview of the representation of women in academic science and engineering at the time the surveys were conducted in 2004 and 2005. A more extensive analysis of changes from 1995-2003, using data primarily from the SDR, can be found in Appendix 2-1, along with an overview of existing research. The committee used many of the themes and issues identified in this research to develop the survey questionnaires, and we hope that the findings presented here—and the many unanswered questions—will form the basis for future research.

OUTLINE OF THE REPORT

The remainder of the report is divided into four topic areas. Chapter 2 presents data on the representation of female faculty in science and engineering as of 2004-2005. The next three chapters present the survey results and analysis, with findings at the end of each chapter. Specifically, Chapter 3 examines the applicant pool for academic positions in research universities and the hiring process. Chapter 4 considers the day-to-day life of academics, examining professional activities, climate, institutional resources (including start-up packages, laboratory space, and access to equipment), and outcomes such as publications, grant funding, and salary. Chapter 5 explores whether there are disparities in the tenure and promotion process in research universities and, if so, whether those disparities are associated with gender. Chapter 6 provides a summary of key findings from the surveys and the committee's recommendations, including questions for future research.

2

Status of Women in Academic Science and Engineering in 2004 and 2005

Over the past 30 years, legislators, government agencies, professional societies, university administrators, and faculty have increasingly endeavored to raise the number of women pursuing higher education and careers in science and engineering (S&E). To a degree, these efforts have succeeded. Women have made substantial strides both in participating in postsecondary S&E education and in attaining careers in the academic workforce.[1] This chapter provides an overview of the representation of women in academic science and engineering at approximately the time of the faculty and departmental surveys (2004 and 2005). In some cases, results from more recent studies have also been included. These data and analyses provide a context for understanding and assessing the results of the surveys, as well as ideas for further research. The findings and recommendations in this report, however, are based solely on the survey data.

The information in this chapter has been compiled from multiple sources. The data are drawn primarily from the Survey of Doctoral Recipients (SDR), conducted every 2 years by the National Science Foundation (NSF), and the National Survey of Postsecondary Faculty (NSOPF), which has been conducted every 5 years since 1988 by the National Center for Education Statistics (NCES) of the Department of Education.[2] The SDR samples all doctoral scientists and engineers, and the present study focuses on the subset who are faculty. The

---

[1] Marschke et al. (2007), write, however, that progress for female faculty has been "glacial" and "excruciatingly slow."

[2] Additional information on the surveys can be found at SRS Survey of Doctoral Recipients at http://www.nsf.gov/statistics/showsrvy.cfm?srvy_CatID=3&srvy_Seri=5, accessed on June 13, 2006; and National Study of Postsecondary Faculty—Overview at http://www.nces.ed.gov/surveys/nsopf/, accessed on June 13, 2006.

NSOPF samples only faculty, and this report concentrates on the subset that is in the natural sciences and engineering. Both NSF and NCES release special reports, which were also consulted.[3]

Data from professional societies were also examined, including the American Association of University Professors (AAUP), which focuses on faculty, and the American Association for the Advancement of Science (AAAS), which surveys its members.[4] In addition, several discipline-oriented societies provided data from member surveys, for example, the Computing Research Association (CRA), the American Mathematical Society (AMS), the American Institute of Physics (AIP), the American Chemical Society (ACS), and the American Society for Engineering Education (ASEE).[5]

Finally, the committee consulted studies conducted by individual universities (e.g., on gender equity, salary, or climate) and publications by individual researchers. An analysis of historical trends in the representation of women in academic science and engineering based on the SDR and NSOPF and a more extensive review of the research literature can be found in Appendix 3-1.

DEGREES EARNED

Evidence of women's representation in science and engineering is often measured first in the attainment of undergraduate and graduate degrees. [6] In 2004, 50.4 percent of all S&E bachelor's degrees went to women.[7] Women received the majority of bachelor's degrees in the agricultural sciences, biological sciences, oceanography, and chemistry, and they were awarded more than 40 percent of the bachelor's degrees in the earth sciences, mathematics and statistics, and atmospheric and other physical sciences, excluding physics.[8]

Of all S&E master's degrees awarded in 2004, 43.6 percent went to women. They received the majority of master's degrees in the agricultural and biological sciences and other physical sciences, excluding physics and astronomy. They were awarded over 40 percent of the master's degrees in the earth sciences and oceanography, mathematics and statistics, and chemistry.[9]

---

[3] See for example NSF (2004b).

[4] For further details on the AAAS surveys, see Chander and Mervis (2001) and Holden (2004).

[5] For further details see Byrum (2001), Ivie et al. (2003), Kirkman et al. (2006), Long (2000, 2002), Marasco (2003), and Vardi et al. (2003).

[6] The percentage of women participating in science and engineering education, however, is lower than the corresponding percentage of women in the U.S. population of 18- to 30-year-olds. See Kristen Olson, *Despite Increases, Women and Minorities Still Underrepresented in Undergraduate and Graduate S&E Education,* NSF Data Brief, January 15, 1999 (NSF 99-320).

[7] Note here S&E is defined as engineering, natural sciences, and the social and behavioral sciences.

[8] Data tabulated by staff, derived from National Science Foundation WebCASPAR database.

[9] Data tabulated by staff, derived from National Science Foundation WebCASPAR database.

In 2005, 37.7 percent of all S&E doctorate degrees went to women. Women were awarded almost 50 percent of Ph.D.s granted in the biological sciences (National Science Foundation, 2006).

FACULTY REPRESENTATION

Despite these encouraging numbers, the number and percentage of women faculty had yet to match these gains. While noticeably increasing throughout S&E disciplines, women continued to be underrepresented among academic faculty relative to the number of women receiving S&E degrees (Nelson and Rogers, 2005). As Table 2-1 shows, in 2003, women comprised between 6 and 29 percent of senior faculty (full and associate professors) in S&E. The largest percentage of full and associate professors was found in the life sciences, while the lowest was in engineering.

Women were more likely to be assistant professors, and as shown in Table 2-2, comprised between 18 and 45 percent of assistant professors in S&E.[10] Again, the largest percentage of female faculty was in the life sciences, and the lowest was in engineering.

These aggregate proportions masked two noteworthy phenomena. First, some departments had greater success in recruiting, retaining, and advancing female faculty than others. Examinations of specific department rosters continued to turn up examples of departments with no female faculty (e.g., Ivie et al., 2003; Nelson and Rogers, 2005).[11] Second, some types of higher education institutions had done better at recruiting, retaining, and advancing female faculty than others. Female science faculty were more likely to be employed by community colleges or institutions that did not offer a doctoral degree, rather than at the large research universities (Nettles et al., 2000; Schneider, 2000). For example, in mathematics in 2005, the percentage of female, full-time, tenured or tenure-track faculty at doctorate-granting institutions was 11 percent; at master's-granting institutions it was 24 percent; and at bachelor's-granting institutions it was 25 percent (Kirkman et al., 2006).

---

[10] Other studies come to similar conclusions. For example, women comprised only 14 percent of all faculty in astronomy in 2003 (Ivie, 2004) and 13 percent of all faculty in physics in 2006 (Dresselhaus, 2007). In mathematics in 2005, only 11 percent of full-time, tenure-track or tenured faculty in doctoral departments were women, while 24 percent of non-tenure-track, full-time faculty were women (Kirkman et al., 2006). In engineering, only 11.3 percent of tenured or tenure-track faculty members were women in 2006 (Gibbons, 2007). It should be noted, though, that over time, these percentages are slowly rising.

[11] In 2006, all of the top 50 chemistry departments had at least one woman on faculty (Marasco, 2006). Continuing the examination of chemistry, for 30 Research I institutions that hired at least five faculty during 1988 and 1997, the percentage of women among hires ranged from 50 percent in one case to zero percent in 8 cases. Some departments hired a greater proportion of women than might be expected in comparison to the proportion of women in the doctoral pool, though in most cases, the proportion of women hired was lower (NAS, NAE, and IOM, 2007).

TABLE 2-1 Science and Engineering Doctorate Holders Employed in Academia as Full-Time Senior Faculty by Sex and Degree Field, 2003

Field	Total (thousands)	Sex		Female (percent)
		Male (thousands)	Female (thousands)	
Natural Sciences	77.5	61.0	16.5	21.3
Physical sciences	17.0	15.3	1.7	10.0
Mathematics	10.2	9.1	1.2	11.8
Computer sciences	2.9	2.4	0.5	17.2
Earth, atmospheric, and ocean sciences	4.3	3.5	0.8	18.6
Life sciences	43.1	30.7	12.4	28.8
Engineering	17.2	16.1	1.1	6.4

SOURCE: Adapted from NSB, 2006.

TABLE 2-2 Science and Engineering Doctorate Holders Employed in Academia as Full-Time Junior Faculty by Sex and Degree Field, 2003

Field	Total (thousands)	Sex		Female (percent)
		Male (thousands)	Female (thousands)	
Natural Sciences	31.6	19.6	11.8	37.3
Physical sciences	5.5	4.3	1.3	23.6
Mathematics	2.8	2.0	0.9	32.1
Computer sciences	1.3	1.0	0.3	23.1
Earth, atmospheric, and ocean sciences	1.8	1.3	0.5	27.8
Life sciences	20.1	11.1	9.0	44.8
Engineering	5.6	4.6	1.0	17.9

SOURCE: Adapted from NSB, 2006.

According to Cataldi et al. (2005:3), "full-time faculty and instructional staff at public doctoral and private not-for-profit doctoral institutions were less likely to be female (32–33 percent) than those at public master's, private not-for-profit baccalaureate, and other institutions (41 percent each), private not-for-profit master's institutions (43 percent), and public associate's institutions." This was a long-standing trend, as noted in NRC's (2001a:155) analysis of NSF data for 1979, 1989, and 1995, which found that women were "least represented among the faculty at Research I and Research II institutions." Summarizing the landscape in an article titled "Where the Elite Teach, It's Still a Man's World," Robin Wilson (2004) wrote, "At the country's big research universities, the vast majority of professors are men."

Related to this is the fact that female faculty tended to be clustered in positions that were part-time, untenured, or at lower ranks. The number of positions off the tenure track—both part- and full-time—had grown dramatically over the past few decades (Anderson, 2002; Bradley, 2004). Comparing full-time to part-time positions, women were less likely to be found in full-time positions. In mathematics, for example, during the fall term of 2005, 37 percent of the part-time faculty at doctorate-granting institutions were women, while only 11 percent of the full-time, tenured and tenure-track faculty were women, and only 24 percent of the full-time, non-tenure-track faculty were women (Kirkman et al., 2006).[12]

Women comprised a particularly small percentage of tenured scientists and engineers in universities and 4-year colleges in 2001 (NSF, 2006). In engineering, for example, the percentage of tenured faculty who were women was 6.2 percent (out of a total of 15,480 faculty). In mathematics and statistics, the percentage was 11.9 percent (of 10,610 faculty), and in the physical sciences, it was 11.1 percent (of 18,930 faculty). In computer and information sciences, the percentage was 17.7 percent (of 2,670 faculty). The biological and agricultural sciences had the highest percentage of tenured faculty who were women, with 21.7 percent (of 30,940 faculty).[13]

Finally, NSF noted in its biennial publication, *Women, Minorities, and Persons with Disabilities in Science and Engineering: 2000* (2000:59), that "within 4-year colleges and universities, female scientists and engineers hold fewer high-ranked positions than do their male counterparts. Women were less likely than men to be full professors and more likely than men to be assistant professors." These findings were confirmed in the 2007 follow-up to that report (NSF, 2007). In a survey of the top 50 departments in several fields, Nelson (2005) found the percentages of women dropped off through the professorial ranks from assistant to associate to full professor in all fields except one.[14] For example, in chemistry, women comprised 21.5 percent of assistant professors, 20.5 percent of associate professors, and 7.6 percent of full professors. In physics, 11.2 percent of assistant professors, 9.8 percent of associate professors, and 4.6 percent of full professors were women. In civil engineering, 22.3 percent of assistant professors, 11.5 percent of associate professors, and 3.5 percent of full professors were women (Nelson and Rogers, 2005).[15]

---

[12] Doctorate-granting institutions are defined as Groups I, II, III, IV, and V. See Kirkman et al. (2006) for complete definitions.

[13] Note these are small gains over 2001 data (compare with NSF, 2003b). The figures here do not agree with those in Table 1-1 due to differences in year of reference, sampling and nonsampling errors, and definitional differences.

[14] The exception was computer science: 10.8 percent of assistant professors, 14.4 percent of associate professors, and 8.3 percent of full professors were women.

[15] Data for chemistry are from 2003; data for physics and civil engineering are from 2002. Newer

Data for faculty at a wider range of institutions were consistent with Nelson's findings (NAS, NAE, and IOM, 2007). For tenured or tenure-track engineering faculty in general in 2005, women comprised 6.3 percent of full professors, 13.2 percent of associate professors, and 19.5 percent of assistant professors (Gibbons, 2007).[16] In physics, women comprised 6 percent of full professors, 14 percent of associate professors, and 17 percent of assistant professors (Dresselhaus, 2007).

The explanation that female faculty on average tended to be younger and so were more likely to be at lower ranks did not completely explain their lower ranks according to the National Research Council (2001a:172), which found "*that at any given career age men are more likely to be in a higher rank* [emphasis in original]." For example, in 1995, in the 10th year since receiving a Ph.D., 8 percent of women and 12 percent of men were full professors; in the 15th year, 33 percent of women and 45 percent of men were full professors; and in the 20th year, 64 percent of women and 73 percent of men were full professors (pp. 172-173). Something other than career age appeared to be causing part of the observed gender differences in rank attainment.

PROFESSIONAL ACTIVITIES AND CLIMATE

In addition to the underrepresentation of female faculty, concerns persisted regarding gender differences in the treatment of faculty. Several studies suggested women were evaluated more harshly and were less likely to be hired into academic positions (Lewin and Duchan, 1971; Steinpreis et al., 1999; Trix and Psenka, 2003; Wenneras and Wold, 1997). The literature also suggested that once hired, women were treated differently than men. Women were less likely to receive tenure or a promotion—the major career milestones for academics—or they spent more time in a lower rank before tenure or a promotion, with negative consequences for their salaries (Long et al., 1993; NRC, 2001; NSF, 2004a). Ginther (2001) found women scientists, in general, were 12 percent less likely than men to be promoted. Long et al. (1993) reached a similar conclusion for women in biochemistry.[17]

Some writers suggested that female faculty received fewer resources than male faculty, with academic salaries being an obvious, much studied, example. Data from the Department of Education revealed that during the 2003 to 2004 academic year, male "faculty with 9/10-month contracts earned an average salary

---

data are available in chemistry. See Marasco (2006) for percentage of female faculty at the nation's top 50 chemistry departments from 2000 to 2006. See NAS, NAE, and IOM (2007) for numbers of male and female faculty in chemistry from 1966-1999.

[16] This is a general trend. According to data collected by the AAUP, about 40 percent of men were full professors, compared to about 20 percent of women. In addition, a greater percentage of women were instructors, lecturers, or had no rank (Curtis, 2004).

[17] Recent data have cast doubt on this position, suggesting significant differences might not occur (Ginther and Kahn, 2006).

of $68,000, and female faculty with contracts of the same length earned an average salary of $55,000" (Knapp et al., 2005). According to an AAUP survey, women's salaries for the academic year 2003 to 2004 continued to remain lower than men's salaries in every category (Curtis, 2005).[18] Curtis explained that women were "still disproportionately found in lower-ranked faculty positions, including non-tenure-track lecturer or unranked positions, which tend to pay lower salaries," and women were "more likely than men to be employed at associate degree and baccalaureate colleges, where salaries are lower" (p. 29). However, studies of salaries of science and engineering faculty, which controlled for such factors as career age, discipline, institution type, rank, and productivity still found disparities in salary (Ginther, 2001, 2004; NRC, 2001b). There was some evidence that the gender gap in academic salaries was shrinking over time (see, for instance, Holden, 2004).

Other resources may not have been equitably held. The 1999 Massachusetts Institute of Technology study (MIT, 1999), for instance, noted women faculty had less laboratory space than men. University departments doled out a variety of resources, including access to research assistants, travel money, lab space and equipment, summer research money, etc.

A third area where inequities were seen to exist was in academic workloads (Fogg, 2003a; Jacobs, 2004; Nettles et al., 2000; Park, 1996). As Park (1996) explained, "Though all university faculty are expected to teach and to serve, as well as to carry out research, male and female faculty exhibit significantly different patterns of research, teaching, and service. Men, as a group, devote a higher portion of their time to research activities, whereas women, as a group, devote a much higher percentage of their time to teaching and service activities than do men" (p. 54). An examination of fall 2003 full-time S&E faculty at Research I institutions in the Department of Education's 2004 NSOPF found that men and women spent, on average, 35.8 percent and 30.3 percent of their time on research activities, respectively. Conversely, women and men spent 46.9 and 41.3 percent of their time on instruction, respectively.[19] Men and women spent almost the same percentage of time on administrative and other activities.[20] Disparities in research

---

[18] Perna's (2002) analysis suggested that female faculty were less likely to receive supplemental earnings, such as from institutional sources or private consulting.

[19] Data were created using the Department of Education's Data Analysis System (DAS), available online at http://www.nces.ed.gov/dasol/. Gender was used as the row variable. The column variables were mean percent time spent on research activities, mean percent time spent on instruction, and mean percent time spent on other unspecified activities. Filters were only Research I institutions, full-time employed, with faculty status, with instructional duties for credit, and with principal fields of teaching as agriculture and home economics, engineering, first-professional health sciences, nursing, other health sciences, biological sciences, physical sciences, mathematics, and computer sciences.

[20] Administrative and other activities are defined as those that occur at the respondent's institution such as administration, professional growth, service, and other activities not related to teaching or research.

time may have had critical consequences, as productivity is the most important component in deciding tenure and promotion cases[21] and in determining salary.

A final area where disparities may have occurred between female and male faculty was in job satisfaction and retention. In general, women were less satisfied in the academic workplace than males (Trower and Chait, 2002), which may have led to unhappiness with one's profession and consequently lower productivity and decreased retention rates. Lawler (1999) noted an additional concern: "unhappiness gets transmitted to younger women starting out and may help scare a new generation away from academia," thus potentially reducing the pool of future academics.

Several studies found women had higher attrition rates than men both prior to and after tenure was granted (August, 2006; August and Waltman, 2004; Carter et al., 2003; Trower and Chait, 2002).[22] Yamagata (2002), for example, found that the attrition rate for female faculty at medical schools was higher than the rate for male faculty from 1980 to 1999 (although the attrition rate for women was decreasing faster than the attrition rate for men and more women were becoming full-time faculty members, resulting in a shrinking gender gap). Johnsrud and Rosser (2002) catalogued a variety of reasons that may explain a faculty member's decision to leave a particular position. These included a variety of individual characteristics, such as personal motivation and satisfaction, as well as institutional support.[23]

Against this backdrop of increasing women's participation in science and engineering but persistent gender gaps, the committee fielded its surveys of faculty and academic departments in 2004 and 2005. Many of the issues and concerns raised by previous data collection and research formed the basis for the survey questions. Again, an analysis of historical trends from 1995 to 2003 and a more extensive review of the literature can be found in Appendix 2-2.

---

[21] As Nettles et al. (2000:8) noted: "Some researchers have argued that most faculty reward systems are based on research performance" (Hansen 1988), and existing research supports this assertion (e.g., Fairweather 1995, 1996; Gomez-Mejia and Balkin 1992; Ferber and Green 1982; Lewis and Becker 1979; Tuckman and Hageman 1976). See also Fairweather (2002).

[22] Although at least one study of 210 departments of computer science conducted in 2002 for the period 1995-2000 found that female faculty had lower turnover than men (Cohoon et al., 2003).

[23] See also Amey (1996).

3

Gender Differences in Academic Hiring

This chapter examines this critical entry point into an academic career—and its components—with a primary focus on differences in hiring outcomes for tenure-track assistant professor and tenured associate or full professor positions, and how these differences might be explained. The following research questions are addressed:

- Is gender associated with the probability of individuals applying for S&E positions in research-intensive institutions?
- Given that an individual applies for a position, does a woman have the same probability of being interviewed as a man?
- Given that an individual is interviewed for a position, does a woman have the same probability of being offered a position as a man?

As the chapter explores the impact of institutional and departmental characteristics, rather than the individual characteristics of potential applicants and job candidates, another way to frame the research questions is, what are the characteristics of research-intensive (Research I or RI) institutions associated with proportionately more applications from women, interviews of women, and offers to women?

The chapter is divided into five sections. We outline the hiring process with a focus on three key parts of the hiring process—applications, interviews, and offers. The final two sections describe faculty perceptions of hiring and institutional policies based on data from our faculty survey. A review of the relevant literature and research and what it suggests we should expect to find in our survey data can be found in Appendix 2-1.

THE HIRING PROCESS

The hiring process consists of a series of decisions made sequentially by an academic department and job applicants. A department is authorized to search to fill a faculty position. The search may be for a senior faculty member who will be offered a tenured position; for a tenure-track position, which has the potential to become a tenured position, but does not provide tenure at the time of hire; or for both. This chapter separately considers tenure-track positions and tenured positions for which the six science and engineering departments in Research I institutions surveyed completed searches in the period 2002-2004. This report does not report on positions off the tenure track because no data were collected on these openings.

This section briefly outlines the steps in the hiring process as follows:

- the department's actions in advertising the availability of a position;
- the individual's decision on whether to apply for the position;
- the department's choice of individuals to interview and to make the first offer to; and
- the individual's choice of whether to accept the offer.

Each of these steps is described below.

Advertising the Position

As part of the process that authorizes a department to fill a faculty position at a tenured or tenure-track level, the department determines the subfield(s) that the individual will be expected to fill (both in a research and teaching capacity). Tenure-track positions at the assistant professor level are advertised nationally in journals and at national conferences. Letters may also be sent to department chairs or faculty in a particular subfield notifying them of open positions. Efforts are generally made to make the hiring process for tenure-track positions appear open and equitable. Advertisements note that the institutions follow Equal Employment Opportunity (EEO) rules, and many ads specifically encourage applications by women and minorities. At this point in the process, it is very likely male and female candidates are equally aware of most positions. That is, there is not likely to be a gender-based information gap.

In addition to national advertising, however, the hiring process for tenure-track positions also involves recruiting that could result in gender differences in application rates. For example, word-of-mouth recruiting practices by faculty may generate differences by gender, intentionally or not, in information about the position available to potential applicants. Search committees may try to overcome the limitations of established networks by making special efforts to increase the number of women applying for a given position.

The recruitment process for tenured positions may differ from the process for tenure-track positions in subtle ways. Although the advertising for tenured positions frequently mirrors the advertising for tenure-track positions, it is also common for a department to formulate a list of the leading candidates, based on its view of who is doing the most interesting and important research in that particular subfield, and to ask those on the list directly if they are interested in applying.

The Decision to Apply

Once a potential applicant is aware of a position, this individual may or may not choose to apply. In making this decision, a potential applicant may receive advice from many people, including the person's mentor, department chair, peers, faculty at various institutions, family members, or spouse. A variety of factors may be taken into account in determining whether to apply. These include expectations about the desirability of the position (salary and benefits, prestige of the department, facilities, or workload); the location; and whether a spouse's or other family member's needs will be met. An important factor may also include the encouragement (or lack thereof) that potential applicants receive from the faculty members that they consult, particularly their dissertation or postdoctoral supervisors.

Requests for Campus Visits, Interviews, and Selection

Once applications arrive, decision making reverts to the institution, typically through an appointed search committee. At this point, the search committee ranks the applicants and determines whom to invite to campus for interviews or for preliminary interviews at professional society meetings. Search committees also consider a variety of factors in determining who they feel are the best candidates, including expectations of future productivity (e.g., research and grants received), ability to meet teaching needs, and perceptions of fit. "Fit" is perhaps the most subjective criterion. It is usually thought of as how well a particular candidate's area of expertise or methodological approach works with the department's current needs or vision for its future strengths and mission. However, it can also focus attention on a candidate's demographic background or personality. Different search committees weigh these factors differently. Top candidates are invited to interview, which usually includes giving a talk about their research. This gives the search committee extra information on a few candidates. At the end of this process, often—but not always—an offer is made to a candidate.

The Decision to Accept or Reject the Offer

The final decision is made by the candidate whether or not to accept the offer. Again, the candidate weighs many factors in making this decision. These include the benefits of the position, other employment opportunities, and the candidate's preferences (possibly also including the preferences of a spouse or family members).

Data on Hiring

Data on the hiring process, as described above, are scant. Unfortunately, nationally representative information is not available. First, there is no national evidence on applicant behavior. It is not known if male and female S&E doctorates apply to positions in a similar manner. Second, evidence of how search committees select one candidate over another is lacking, perhaps because the selection process can be difficult to quantify. Third, there is little evidence describing the number of individuals who go through the hiring process. While departments collect information on the number of applicants who apply for a position and are interviewed, and while gender is often noted for these individuals, data are rarely made public for rather good reasons, including the right to privacy of job applicants.[1] Further, comparable data on hiring activities at different universities are not generally available to allow an examination of how university and departmental search policies and practices affect hiring outcomes. National statistics such as the National Survey of Postsecondary Faculty or the Survey of Doctorate Recipients focus on individuals in their current positions. The SDR asks doctorates about their postgraduate plans and whether they are interested in a postdoctoral or academic position, but does not follow respondents any further. As a result, this chapter will draw primarily from this study's departmental survey described in Chapter 1 and in Appendix 1-4.[2]

The survey asked chairs of the six targeted departments in each of the Research I institutions to report whether they had conducted any searches during the 2002-2003 or 2003-2004 academic years. Of the 492 surveyed, 417 responding departments reported a total of 1,218 searches, ranging between 1 and 15 searches per department. Responding departments were asked to identify whether the search was for a tenured or tenure-track position. In a few instances respondents wrote in "both" (17 out of 1,218), and to a lesser degree "target of opportunity" (5 out of 1,218). A few (40 out of 1,218) left this question unanswered. Respon-

---

[1] However, some institutions do release their analyses of hiring. An excellent example is the 2003 gender equity report undertaken at the University of Pennsylvania, which presents important data for consideration and evaluation while maintaining anonymity. See http://www.upenn.edu/almanac/v50/n16/gender_equity.html. See also the report, *University of California: Some Campuses and Academic Departments Need to Take Additional Steps to Resolve Gender Disparities among Professors*, Report by the California State Auditor, 2001, available at http://www.bsa.ca.gov/pdfs/reports/2000-131.pdf. See also the report by the Commission on the Status of Women at Columbia University, *Advancement of Women Through the Academic Ranks of the Columbia University Graduate School of Arts and Sciences: Where Are the Leaks in the Pipeline?*, available at http://www.columbia.edu/cu/senate/annual_reports/01-02/Pipeline2a_as_dist.doc.pdf.

[2] The committee acknowledges that the *p*-values for all the data presented are unadjusted and that many of the data presented are interconnected.

dents were then asked to provide data on the number of applicants and interviewees for each advertised position by gender. Finally, they were asked to identify the gender of the individual who was first offered the position and the gender of the person who was ultimately hired.[3]

In general, departments were much more knowledgeable about the later stages of the hiring process and thus provided more complete data on offers and hires than on interviews or applicants. The number of cases for which we had complete information on applicants, interviewees, first offers, and hires—all disaggregated by gender—varied between 534 cases (with complete hire information) and 758 cases (with complete applicant information). Thus, the number of cases considered in this chapter depends on the stage of the hiring process. Only tenured and tenure-track cases are considered in the analysis. For each stage in the hiring process (applications, interviews, offers), descriptive statistics based on the data collected from the departmental survey are first presented. Then, the appropriate statistical models are fit in order to understand the departmental characteristics associated with the percent of females at each stage of the hiring process.

APPLICATIONS FOR FACULTY POSITIONS

A necessary precondition for hiring a female faculty member is to have women who are interested in applying for the position. The survey data clearly show that some departments are more successful than others in attracting female applicants.[4] Moreover, our data show that there are still a number of positions for which no women apply.

Throughout this report, we will present summary statistics, such as the following ones, that state current values for men and women across the six disciplines surveyed. These statistics do not reflect the survey weights[5] and are not treated for the different degrees of nonresponse that depended on the characteristic examined. Therefore, these statistics are NOT appropriate estimates of any national characteristics for men and women, but instead are quick impressions of the data collected, which are often the beginning of a more meaningful analysis that is conditional on the disciplinary area.[6]

---

[3] A limitation of the survey was that it did not ask for the gender of every candidate offered a particular position.

[4] Note that this analysis implies nothing about the quality of applicants. Some people apply for jobs for which they are not a very good fit. The committee did not assess whether male and female applicants would behave any differently in this regard.

[5] Recall that the committee's survey was stratified in order to collect similar numbers of respondents in each of the six disciplinary areas, and therefore respondents from different disciplines have different survey weights.

[6] These estimates would be useful as national estimates only in situations in which the disciplines are relatively homogeneous with respect to a given characteristic and the nonresponse which occurred was such that nonrespondents did not differ in their characteristics from respondents.

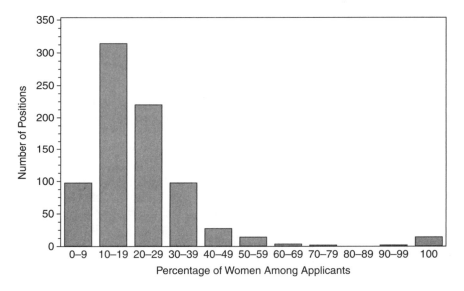

FIGURE 3-1(a) Percentage of females among applicants to all tenured and tenure-track positions.

Descriptive Data

While women are increasingly receiving Ph.D.s in Science and Engineering (S&E), they are still greatly outnumbered by men in terms of applications for Research I positions. For tenure-track jobs, the median number of applications a department receives is 52 applications from men and 8 applications from women—or about 7 applications from men for every application from a woman. For tenured positions, the median number of applications a department receives is 40 applications from men and 8 from women, for a ratio of 5 to 1.[7] Figure 3-1(a) presents a histogram of the percentage of female applicants for all positions; Figure 3-1(b) presents this information for tenured positions; and Figure 3-1(c) presents this information for tenure-track positions.

Overall, departments received from 1 to 800 applications for their advertised tenure-track positions ($n = 626$), and 1 to 500 applications for tenured positions ($n = 128$). Departments recorded only 1 applicant for 17 (3 percent) tenure-track positions and 9 (8 percent) tenured positions. The survey results showed that 3 men and 2 women were hired through "target of opportunity" positions where

---

[7] These figures are medians. The median was used because the data are skewed; there are a few positions that had hundreds of applicants. The mean number of applications for tenure-track jobs was 85 applications from men and 17 from women. The mean number of applications for tenured jobs was 78 from men and 17 from women.

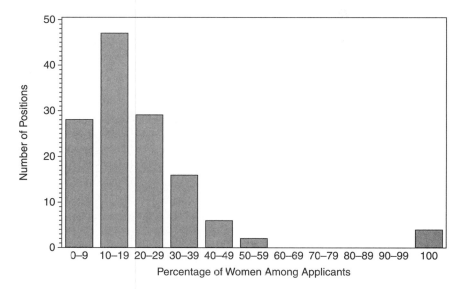

FIGURE 3-1(b) Percentage of women among applicants to all tenured positions.

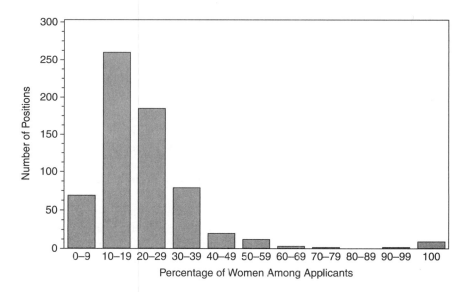

FIGURE 3-1(c) Percentage of women among applicants to all tenure-track positions.
SOURCE: Survey of departments carried out by the Committee on Gender Differences in Careers of Science, Engineering, and Mathematics Faculty.

TABLE 3-1 Number of Tenured and Tenure-Track Positions with Complete Information About the Gender of Applicants by Discipline

Discipline	Tenured	Tenure-Track
Biology	24 (15)	118 (43)
Chemistry	19 (16)	128 (47)
Civil Engineering	13 (9)	73 (33)
Electrical Engineering	14 (9)	75 (27)
Mathematics	31 (16)	98 (37)
Physics	27 (14)	134 (47)

NOTE: Numbers in parentheses are the numbers of separate departments offering those positions.
SOURCE: Survey of departments carried out by the Committee on Gender Differences in Careers of Science, Engineering, and Mathematics Faculty.

the position by intention was offered to only 1 candidate, though the rank at hire was not known. Table 3-1 shows the number of cases with complete applicant information by discipline and type of position (tenured or tenure-track). Note that the number of cases across discipline and type of position combinations is roughly similar, so no discipline contributes an inordinate proportion of the data to the analyses that follow.

Another finding is that for job openings for which only individuals of 1 gender applied, that gender was more likely to be male. For tenure-track positions, there were only 9 openings for which no men applied (only women applied), and 8 of these were cases in which only 1 woman applied. On the other hand, there were no female applicants (only men applied) for 32 tenure-track positions, or about 6 percent of available positions, with only 9 of these positions having a single applicant. Similar findings were seen for tenured positions. For 2 positions, no men applied. These were the 2 cases in which there was only 1 applicant. Conversely, no women applied to 16 tenured jobs, or 16.5 percent of the positions; only 7 of these were single-applicant positions. This finding may lend credence to the anecdotal argument sometimes propounded by chairs or search committees that no women applied for particular advertised positions (Brennan, 1996; see especially p. 9).

Considering the data by discipline, in the instance of tenure-track positions, most of the cases (29 of 32) in which only men applied occurred in physics or the engineering fields. For tenured positions, 10 of the 16 cases occurred in chemistry (6) and physics (4). This may reflect the fact that engineering and physics have a lower percentage of female doctorates or that female engineers and physicists are more likely to prefer employment outside of major research universities.

Finally, how do the percentages of female applicants relate to the percentage of women in the doctoral pool from which departments are drawing? One might expect the proportion of female applicants to be similar to the percentage

TABLE 3-2 Percentage of Women in the Doctoral Pool and Distribution of the Percentage of Women among Job Applicants for Tenure-Track Positions by Discipline

Discipline	1999-2003 All Doctorate-Granting Institutions (percent)	1999-2003 Research I Institutions Only (percent)	Mean Percentage of Female Applicants for Tenure-Track Positions (percent)[a]
Biology	45	45	26 (8, 25, 50)
Chemistry	32	32	18 (6, 15, 39)
Civil Engineering	18	18	16 (0, 10, 100)
Electrical Engineering	12	12	11 (0, 10, 22)
Mathematics	27	25	20 (9, 20, 34)
Physics	15	14	13 (0, 10, 27)

NOTES: In parentheses, we show the 5th percentile, the median, and the 95th percentile (computed over all tenure-track positions in each discipline) of the percentage of females among applicants. Only those tenure-track positions with complete information about the gender of candidates were included in these calculations (as in Table 3-1).

[a] Mean percentage of female applicants computed as the average (over all tenure-track positions) of the percentage of females in the applicant pool.

SOURCE: Ph.D. data are from the National Science Foundation. WebCASP distribution of the percentage of female applicants was computed using the same data used to construct Table 3-1.

of doctorates awarded to women in S&E across each of the disciplines. Table 3-2 suggests that this relationship is more complex. In the table, the second column shows percentages of doctorates awarded to women in the period 1999-2003 by doctorate-granting institutions, while the third column shows percentages of Ph.D.s awarded to women by the subset of Research I institutions.[8] Data on the proportion of women among all applicants for tenure-track jobs by discipline are presented in column four.

In examining Table 3-2, it is important to note that while the second and third columns reflect averages over individuals, the last column relates to the percentage of women averaged over job openings. Thus, the values are not strictly comparable. An individual can apply to more than one job and may be counted multiple times as an applicant. If women are more likely to apply to multiple jobs than men, then the percentage of women among applicants is overestimated. Conversely, if women only apply to a few positions while men apply to many, then the average percentage of women applicants (and the rest of the distribution of the percentage of female applicants) is underestimated.

Table 3-2 shows that the percentage of applications from women are

---

[8] For a discussion of how to define the "pool of qualified candidates," see NAS, NAE, and IOM (2007).

consistently lower than the percentage of Ph.D.s awarded to women. There are, however, substantial differences among the disciplines in how much they are lower. In electrical engineering, mathematics, and physics, the percentage of women applying for faculty positions is only modestly lower than the percentage of women receiving Ph.D.s. However, in the fields with the largest representation of women with Ph.D.s—biology and chemistry—the percentage of Ph.D.s awarded to women exceeds the percentage of applications from women by a large amount. This finding should be further explored. Possible explanations that might be tested in follow-on research include:

- Female biology and chemistry doctorates prefer occupations outside of research-intensive institutions relative to men (for example, in higher education, but in liberal arts colleges; in education as K-12 teachers; or in industry or government);
- As the percentage of doctorates awarded to women increases, departments may make fewer special efforts to encourage women to apply for faculty positions; or
- Female Ph.D.s in biology and chemistry apply for fewer jobs than women in other fields relative to men.

The first hypothesis may also, to a greater or lesser extent, hold for the smaller disparities found in civil engineering, electrical engineering, mathematics, and physics.

Another study examining the percentage of women in Ph.D. pools relative to the percentage of female faculty also found mixed results (NAS, NAE, and IOM, 2007). Comparing data for faculty who were tenure-track or tenured in 2003 with earlier averages of doctorates revealed that in engineering, chemistry, and the physical sciences, there was a smaller percentage of women in the Ph.D. pool than in assistant professor positions, while in the life sciences, computer sciences, and mathematics, the percentage of women in the pool of doctorates was larger. Comparing the doctoral pool to associate professors in engineering and life sciences, the percentage of women in the pool exceeded the percentage of female associate professors. In computer science, chemistry, the physical sciences, and mathematics, there was a greater percentage of female associate professors. Considering full professors, the percentage of female full professors in most fields was smaller than the percentage of women in the relevant doctoral pool.

Statistical Analysis

Having summarized earlier in this chapter the literature on the factors that are potentially associated with the percentage of applicants who are women, we now investigate whether the data on hiring collected in our surveys support the hypotheses put forth by earlier investigators. In our applicant models, the following institutional, departmental, and position-level variables measured in our survey were used as explanatory variables: discipline, type of position (tenured, tenure-track), whether the institution is private or public, the prestige level of the department advertising the position, the proportion of females in the search committee, the number of family-friendly policies advertised by the institution, whether the search committee chair is a man or a woman, the percentage of female faculty in the department, and the size of the metropolitan area in which the institution is located.

We first investigated whether any of these factors are associated with the probability that no women apply to a position.[9] To do so, we first created a binary variable with the value 0 if there were no female applicants and the value 1 if at least one woman applied to the position. We excluded for this analysis those positions identified as target of opportunity and open rank positions. We fitted a logistic regression model to the binary outcome variable and included as predictors in the model the institutional, departmental, and position-level variables listed above, as well as two-way interactions between discipline and the other predictors to investigate whether any of the potential effects of predictors is discipline-dependent. To account for possible correlations within positions advertised by the same institution, we implemented the method of generalized estimating equations (GEE) to compute standard errors for all parameter estimates that account for possible correlations across positions in the same institution.

We found the probability that at least one woman would apply to a position is associated with the set of discipline indicators ($p = 0.03$), type of position ($p < 0.0001$), type of institution ($p = 0.08$), prestige of the institution ($p = 0.04$), and the number of family-friendly policies in effect at the institution ($p = 0.001$). No other factor was statistically associated with the probability of at least one female applicant. Results can be more easily understood by looking at the adjusted means of the differences in the probability of no female applicant across levels of some of the statistically significant factors. These adjusted means are the means computed after "adjusting for" or "accounting for" all other effects in the model. Technical details and the tables are given in Appendix 3-2. We then focused on all positions and modeled the number of female applicants as a function of the same independent variables listed above. To do so, we fitted a Poisson regression model to the number of female applicants and used total number of applicants as an exposure variable. Possible correlation across positions advertised by the same institution was accounted for when computing standard errors of parameter estimates via the method of generalized estimating equations method. Again, we only included positions that were advertised as tenured or tenure-track.

As expected, we found statistically significant differences across disciplines in the proportion of females in the applicant pool. Biology, chemistry, and math-

---

[9] The vast majority of both tenure-track (94 percent) and tenured (83.5 percent) positions had at least one female applicant.

ematics had significantly higher proportions of female applicants than did all other disciplines across all types of institutions and positions. The proportion of female applicants in civil engineering, physics, and electrical engineering was significantly lower. The type of position was not substantially associated with the proportion of females in the applicant pool. The percentage of females among applicants to tenured positions was similar to the percentage of females among applicants to tenure-track positions.

It has been speculated that the appearance of a women-friendly environment attracts female applicants. Our results confirm this view. The percentage of women in the search committee and whether a woman chaired the committee were both significantly and positively associated with the percentage of women in the applicant pool ($p = 0.01$ and $p = 0.02$, respectively). For every 1 percent increase in the percentage of females in the search committee, we can anticipate an increase of about 0.7 percent in the percentage of women in the applicant pool. In contrast, the number of family-friendly policies advertised by the institution did not appear to be associated with the percentage of female applicants. Other factors including type of institution (public or private), prestige of the institution, and location of the institution had no association with the percentage of women in the applicant pool.

These results may thus support the argument that an individual applicant's characteristics are relatively more important in determining application behavior. Institutions wishing to increase the number of applications from women may have to rethink current efforts or consider new strategies.

SELECTION FOR INTERVIEWS FOR S&E JOBS

This section examines the representation of women among candidates whom departments choose to interview. Prior to this survey, few data were available about the probability that a female applicant for an academic position will be interviewed as compared with the probability that a male applicant will be interviewed. There is, however, substantial literature suggesting that reviewers tend to discount the credentials and qualifications of female job applicants. Insofar as this discounting occurs among academic searches such literature might be relevant.

The committee's departmental survey allows an examination of the percentage of women being interviewed and offered positions. This section examines the interviewing behavior of departments.

Descriptive Data

Our survey data allowed us to examine the actual behavior of departments for the 545 tenure-track and 97 tenured openings for which we have gender data for applicants, interviewees, offers, and ultimate hires. Across all the positions—tenure-track or tenured—an average of four men and one woman were interviewed for any particular position. A cynical reader might wonder if this is the case because search committees are attempting to show they are fulfilling a diversity mandate by interviewing a woman. However, an examination of the data on the percentage of women interviewed reveals that the percentage does not decline as the number of interviews undertaken increases, as it would if each job search interviewed only one woman for appearances' sake. This finding, however, masks two other important findings.

First, our survey data allowed us to examine the actual behavior of departments for the 545 tenure-track and 97 tenured openings for which we have gender data for applicants, interviewees, offers, and ultimate hires. The second and fourth columns of Table 3-3 draw on information from Table 3-2; that is, the mean percentage of female applicants for tenure-track jobs and the mean percentage of female applicants for tenured jobs. The third and fifth columns present the mean percentage of female interviewees for tenure-track positions and for tenured positions.

As the table shows, in every instance, the mean percentage of female interviews exceeds the mean percentage of applications from women. With the exception of civil engineering, for which the median percentage of female interviewees for tenured positions is zero, results are similar if we compare median percentages (rather than mean percentages), but we do not show those here. (The reason for a zero percent median percentage of women in interview pools in the case of civil engineering is the small sample size of 12 cases.)

Even though the percentage of females in interview pools exceeds the percentage one might expect from the representation of women in applicant pools, no woman was interviewed for 155 (28 percent) tenure-track positions and 41 (42 percent) tenured jobs. Of course, part of this number is comprised of cases for which there were no female applicants. Still, in 124 tenure-track job openings (23 percent), at least 1 woman applied, yet no women were interviewed. In 23 (24 percent) tenured jobs, at least 1 woman applied, but no women were interviewed. These figures are substantially higher than for men. No men were interviewed for 18 tenure-track positions or 3 percent (in nine of those cases, there were no male applicants) and for 4 tenured positions or 4 percent (in 2 of those cases, there were no male applicants).

Table 3-4 shows that for tenure-track jobs, mathematics by far had the lowest proportion of positions for which no women interviewed, followed by biology and chemistry. (These proportions are computed using all cases, including those with no female applicants.) For tenured positions, biology had the lowest proportion of positions for which no women interviewed, followed by physics.

At first glance, the proportion of positions for which no women were interviewed for tenure-track positions might seem high. In all cases, however, the percentage of positions for which no women interviewed was below what might have been expected if gender played no role in the process of selection of interview candidates and if we assume qualifications are not gender-dependent. For

TABLE 3-3 Mean Percentage of Females Among Applicants and Among Interviewees in Each Discipline

Discipline	Tenure-Track			Tenured		
	Applicants (%)	Interviewees (%)	n	Applicants (%)	Interviewees (%)	n
Biology	25	30	111	29	34	20
Chemistry	18	25	123	25	30	18
Civil Engineering	16	30	72	19	30	13
Electrical Engineering	11	19	75	9	17	12
Mathematics	20	28	96	15	21	28
Physics	13	20	124	19	32	25

NOTES: Means were computed as the average (across all positions with complete information about gender of applicants and interviewees) of the percentage of females among applicants (or interviewees) for the position.

The numbers of positions listed in this table are smaller than the numbers listed in Table 3-1, and the mean percentage of female applicants to tenure-track positions are different from those displayed in Table 3-2. This is because here we only considered positions for which complete gender information about all applicants *and* interviewees was available.

SOURCE: Survey of departments carried out by the Committee on Gender Differences in Careers of Science, Engineering, and Mathematics Faculty.

TABLE 3-4 Percentage of Positions for Which No Women Were Interviewed by Type of Position

Discipline	Tenured		Tenure-Track	
	Actual Percentage of All-Male Interview Pools	Probability of All-Male Pools[a]	Actual Percentage of All-Male Interview Pools	Probability of All-Male Pools[a]
Biology	25 (20)	18	22 (111)	24
Chemistry	50 (18)	24	22 (123)	37
Civil Engineering	46 (13)	35	33 (72)	42
Electrical Engineering	42 (12)	62	35 (75)	56
Mathematics	39 (28)	44	13 (96)	33
Physics	32 (25)	35	38 (124)	50

NOTES: Actual number of cases is given in parentheses. The expected number of positions with no women interviewed given the size and gender composition of the applicant pools (see Table 3-3) is computed as described in the text.

The percentage of positions for which no women were interviewed is based on tenured and tenure-track positions for which complete information about gender of all interviewees was available. The data used to construct these values are the same as those used to calculate the statistics showing those interviewed divided by the total number of positions of each type and in each discipline for which complete gender information for all interviewees was available.

[a]These values are the probabilities of an all-male interview pool assuming that five interviewees were selected, the population of applicants was very large, and the frequency of men and women in the applicant pool equaled the percentages from Table 3-3.

SOURCE: Survey of departments carried out by the Committee on Gender Differences in Careers of Science, Engineering, and Mathematics Faculty.

example, assuming five candidates were interviewed for each position, using a simple binary calculation and the proportion of females in the applicant pool from Table 3-3, for tenure-track positions we would expect about 50 percent of the interview pools to include no women in physics, 56 percent in electrical engineering, and 42 percent in civil engineering—the three areas with the lowest representation of women among applicants. In biology, we would expect about 24 percent of the tenure-track interview pools to include no women, again assuming five individuals are on average interviewed for each tenure-track position. In chemistry, the expected percentage of interview pools with no women is 37 percent and in mathematics it is 33 percent. In all cases the percentage of male-only interview pools for tenure-track positions in the six disciplines is smaller than the corresponding probability of an all-male pool. There are significant discipline differences. Electrical engineering and mathematics have the largest difference (21 percent and 20 percent, respectively) between their probability of an all-male pool and their actual interview pools of applicants.

This finding suggests that once tenure-track women apply to a position, departments are on average inviting more women to interview than would be expected if gender were not a factor, or women who apply to tenure-track or tenured positions in research-intensive institutions are, on average, well qualified. It is important to note that these higher rates of success do not imply favoritism, but may be explained by the possibility that only the strongest female candidates applied for Research I positions. This self-selection by female candidates would be consistent with the lower rates of application by women to these positions.

For tenured positions, the expected percentage of interview pools with no women are 18, 24, 35, 62, 44, and 35 percent for biology, chemistry, civil engineering, electrical engineering, mathematics, and physics, respectively. The situation for tenured positions is much less clear. Electrical engineering, mathematics, and physics have smaller all-male interview pools than their probability pools. This is particularly true for electrical engineering, which had male-only interview pools 42 percent of the time compared to a probability of 62 percent. However, civil engineering, chemistry, and biology had larger all-male interview pools than expected, with chemistry being the most notable. Fifty percent of the interview pools for tenured positions in chemistry were all-male, while the probability value was 24 percent. This finding highlights the importance of disaggregating survey data by discipline.

Factors Associated with a Higher Percentage of Female Interviews

As with the analysis of applications, the analysis of interviews focused on departmental and institutional variables. Most of the factors in the applicant model are also used here: discipline; departmental climate, as measured by female faculty; female faculty on the search committee and family-friendly policies; public versus private universities; and prestige. Much of the literature on making hiring more equitable focuses on bringing actors with a broader view from outside the department into the decision making, so we expect intervention by a dean might also be positively related to the probability of interviewing a woman.

Because departments draw from the pool of applicants in deciding whom to interview, this analysis controls for the percentage of applications from women— the dependent variable from the last model. We expect a positive relationship between the percentage of applications from women and the percentage of interviewees who are women.

Statistical Analysis

The percentage of women in the interview pool appears to exceed the percentage of female applicants in all areas. We now investigate whether the percentage of women in the interview pool is associated with the institutional, departmental, and position-level characteristics described earlier and with two additional predictors: the percentage of female applicants and an indicator of whether the composition of the interview pool is reviewed by a dean or other committee external to the search committee. We proceeded as we did when analyzing the percentage of female applicants. We first fitted a logistic regression model to the probability of no women in the interview pool. We then considered all positions and fitted a Poisson multiple regression model to the number of women in the interview pool to investigate whether institutional or position-level attributes are associated with the representation of women in the interview pool. We used the size of the interview pool as an exposure in the model, since the range in interview pool size was quite large, from 1 to 22. (The mean number of candidates interviewed for a position was 5.) In both cases, we accounted for the possible correlation among positions advertised by the same institution by computing standard errors of parameter estimates using the GEE method. The total number of cases considered for these analyses was 667. Of the 667 cases, there were no women in interview pools in 188 cases.

We have argued earlier that the probability of no women in interview pools is below what might be expected across many of the disciplines we reviewed. Results from the logistic regression modeling suggest further that the probability of female interviewees increases when the percentage of female applicants increases, as would be expected ($p < 0.0001$), with the percentage of women in the search committee (borderline significant, $p = 0.06$) and with the number of family-friendly policies advertised by the university (borderline significant, $p = 0.07$). When we account for all covariates, the adjusted mean probability that a woman who has applied to a position receives an invitation to interview is lowest in biology and not significantly different in any of the other disciplines. This would be expected given that biology has significantly more female applicants than other disciplines. The probability of women in the interview pool is significantly lower when the position is advertised as tenured than when it is advertised as tenure-track (p-value $= 0.013$). No other factor was significantly associated with the probability of having at least one woman in the interview pool.

Adjusted means of the probability of at least one woman in the interview pool, with the corresponding 95 percent confidence interval for the true mean probability, are presented in the table in Appendix 3-4. The values in the table corresponding to differences between levels of an effect represent the ratio of the odds ratios in each of the two levels. For example, if the probability that a woman will be interviewed in biology is 0.51, the odds ratio 0.51/0.49 is 1.04, meaning a female applicant is 4 percent more likely to be interviewed than not. If for chemistry the corresponding odds ratio is 4 (0.8/0.2, according to Appendix 3-4) then the ratio of odds ratios between biology and chemistry is 1.04/4 = 0.26. In other words, the "advantage" of a female applicant in biology is only 26 percent of that of a female applicant in chemistry. Calculation of all standard errors (and consequently, confidence intervals) in the table in Appendix 3-4 required using the Delta method. (The Delta method is described in Appendix 3-7.)

When we focused on the number of women in each interview pool, we found that the percentage of female applicants is significantly (and positively) associated with the percentage of females in the applicant pool ($p < 0.0001$) and varies across discipline. For every 1 percent increase in the proportion of female applicants, the proportion of female interviewees increased by approximately 2 percent. The proportion of women in the interview pool was significantly lower in biology, electrical engineering, and physics relative to the other three areas. The effect of discipline, however, is difficult to interpret since the interaction between discipline and other factors is statistically significant. For example, the proportion of women interviewed in mathematics was not the same at public or private institutions. The difference in the percentage of female applicants between mathematics and civil engineering was larger in private institutions. Furthermore, women appear to be interviewed at a higher rate in the top 10 electrical engineering departments than in electrical engineering departments with lesser prestige. Because interpretation of main effects is problematic when interactions are present, we do not present adjusted means resulting from this analysis. No factor other than discipline and the representation of women among applicants (plus some interactions) was found to be associated with the percentage of women in interview pools.

OFFERS MADE

The final step in the search process is making a offer to one of the individuals interviewed. This section examines the percentage of offers made to women and the factors that may have an impact on this percentage. Table 3-5 presents data on whether the department's search results in a first offer to a woman or a man, for the 108 tenured and 583 tenure-track jobs for which we have information on the gender of the applicant to whom an offer was made.

As the table illustrates, women received the first offer about 29 percent of the time for tenure-track positions and 31 percent of the time for tenured positions.

In Table 3-6, we present the distribution, over departments, of the percentage of women interviewees and offers for tenure-track and tenured jobs, which dem-

TABLE 3-5 Percent of First Offers by Gender and Type of Position

Type of Position	First Offer to a		
	Female	Male	Total
Tenured	31	69	108
Tenure-track	29	71	583

NOTES: Only those positions for which complete gender information about interviewees to whom the first offer was extended are included. Thus, the total number of positions on which this table is based is smaller than the numbers shown in Table 3-4. These percentages represent offers in all six disciplines, and therefore may hide important disciplinary differences.

SOURCE: Survey of departments carried out by the Committee on Gender Differences in Careers of Science, Engineering, and Mathematics Faculty.

onstrates that there is variability by discipline hidden by Table 3-5. However, the general pattern remains. Once again—similar to the case for interviews compared to applicants—women receive a greater percentage of first offers than interviews for all fields in the case of tenure-track positions. This finding also holds for tenured positions, except—interestingly—for biology.

Factors Associated with a Higher Probability that a Woman Will Be Offered a Position

The department typically decides who will receive an offer. Thus, the statistical analysis of offers made focused on departmental and institutional variables. Most of the factors included in the applicant and interview models are also used here: discipline; departmental climate, as measured by female faculty, female faculty on the search committee and family-friendly policies; public versus private universities; prestige; and intervention by a dean in the selection process. For availability, the model for offers uses the percentage of interviewees who were women—the dependent variable from the last model. It is assumed that there is a positive relationship between the percentage of interviews of women and the likelihood a woman will be offered the position.

Statistical Analysis

The response variable of interest was binary: a woman was first offered the position or the position was offered to a man. We considered all the institutional and position-level variables described earlier, with the following modifications. Instead of the percentage of female applicants, we now included the percentage of women in the interview pool, and instead of an indicator of whether the candidate pool is reviewed by a dean or an external committee, we included an indicator of whether a dean approves the hiring recommendation made by the committee. Since the probability that a woman will be offered the position when none was interviewed is clearly zero, we restricted these analyses to those positions for which interview pools included at least one woman. Similarly, we also deleted from these analyses those positions for which all interviewees were women. Thus, results presented here are *conditional* on having at least one woman and at least one man in the interview pool.

The only two factors that appear to be associated with the probability that a woman will be offered the position first are the percentage of women in the interview pool ($p < 0.001$) and whether the dean approved an offer (weak association with $p = 0.06$). When the dean reviews offers, the probability that a woman will be offered a position is 0.38, with a confidence interval of 0.26 to 0.50. This value is significantly larger than the 0.06 (95 percent confidence interval of 0.00 to 0.51) obtained in cases in which the dean has no role in reviewing offers. (The uncertainty around this latter value is high because of a very small sample size.

TABLE 3-6 Distribution of Percentage of Interviews with and Offers, to Women by Discipline

Discipline and Type	Distribution of Percentage of Females in Interview Pool					Offers to Females	Total
	0-20 percent	21-40 percent	41-60 percent	61-80 percent	81-100 percent		
Biology							
Tenured	0 (4)	3 (10)	1 (3)	0 (0)	2 (2)	6	19
Tenure-track	3 (36)	15 (44)	10 (15)	6 (8)	1 (1)	35	104
Chemistry							
Tenured	0 (9)	2 (3)	1 (3)	0 (0)	1 (1)	4	16
Tenure-track	6 (50)	24 (53)	7 (14)	0 (2)	0 (0)	37	119
Civil Engineering							
Tenured	0 (6)	0 (1)	1 (1)	2 (3)	0 (0)	3	11
Tenure-track	3 (33)	9 (21)	3 (7)	2 (4)	7 (7)	24	72
Electrical Engineering							
Tenured	1 (6)	1 (2)	1 (2)	0 (0)	0 (0)	3	10
Tenure-track	3 (38)	18 (31)	1 (3)	1 (1)	0 (0)	23	73
Mathematics							
Tenured	2 (17)	6 (9)	0 (0)	0 (0)	1 (1)	9	27
Tenure-track	9 (42)	11 (34)	4 (12)	1 (2)	4 (4)	29	94
Physics							
Tenured	0 (12)	4 (7)	0 (1)	1 (2)	3 (3)	8	25
Tenure-track	7 (83)	4 (21)	3 (9)	2 (2)	5 (6)	21	121

NOTES: Numbers in parentheses correspond to the total number of positions in each group defined by discipline, type, and percentage of women in the interview pool. The next-to-last column shows the total number of positions by discipline and type offered to a woman. The last column shows the total number of positions by discipline and type for which we have complete information on the gender of the candidate receiving the first offer.
SOURCE: Survey of departments carried out by the Committee on Gender Differences in Careers of Science, Engineering, and Mathematics Faculty.

In almost all cases, deans play a role at the time of offering a tenure-track or tenured position to an applicant.) The size of the "dean effect" must therefore be interpreted cautiously. For every 1 percent increase in the percentage of females in the interview pool, the probability that a woman would be offered the position increased by about 5 percent. Finally, the probability that a woman would be offered the position was lowest at the top 20 research-intensive institutions compared with non-top 20 research-intensive institutions surveyed. At the highest prestige institutions (top 10), the probability that a woman would get an offer approached significance ($p = 0.08$). No other factors were associated with the probability that a woman would get an offer.

HIRES

Explaining hires made is more difficult, as the decision to hire involves the department, which makes the offer, and the applicant, who accepts. The committee's departmental survey does not have information on characteristics of those ultimately hired, beyond their gender. However, the committee's faculty survey did ask faculty some questions about reasons for accepting the position offered to them. Answers to these questions are explored in the next section of this chapter.

Table 3-7 presents data on the gender of the individual receiving the first offer and the gender of the faculty member ultimately hired for tenure-track positions.

In 95 percent of the cases in which a man was the first choice for a position, a man was ultimately hired in that position. Compare this to the case for women, where only 70 percent of cases in which a woman was first offered a position was a woman ultimately hired. In 30 percent of the cases in which women were offered first, a man ultimately ended up in the position.[10]

Table 3-8 presents data on the gender of the individual receiving the first offer and the gender of the faculty member ultimately hired for tenured positions.

In all cases in which a man was offered the position first, a man was ultimately hired. In only 77 percent of the cases in which a woman was offered the position first was a woman ultimately hired. In 23 percent of the cases in which a woman was offered the position first, a man was ultimately hired, again suggesting that if the woman who is first offered the position does not accept, there is a substantial chance the job will go to a man.

---

[10] Note, however, that we do not know if the person first offered and the person hired are the same person, where the genders are the same. Nor do we know how many offers were made before someone was eventually hired. Since men outnumber women in the offers made, one would expect that the proportion of times women turn down an offer, resulting in a man being ultimately hired, should be higher than the proportion of times that men turn down an offer, resulting in a woman ultimately being hired.

TABLE 3-7 Percent of Candidates of Each Gender Who Received the
First Offer and Gender of Candidates Who Eventually Accepted Each
Tenure-Track Position

Position Was Offered to	Person Hired Was a	
	Female	Male
Female	70 (107)	30 (46)
Male	5 (19)	95 (362)

NOTES: Number of cases is given in parentheses.
Table 3-7 is based on the subset of the positions used to construct Table 3-6 for which the
gender of the person who accepted the position was known. We do not know from these data
whether the person who accepted the position is the same person who received the first offer,
even in those cases in which the gender is the same.
SOURCE: Survey of departments carried out by the Committee on Gender Differences in
Careers of Science, Engineering, and Mathematics Faculty.

TABLE 3-8 First Offer and Person Hired for Tenured Position, Percent
by Gender

Position Was Offered to	Person Hired Was a	
	Female	Male
Female	77 (20)	23 (6)
Male	0 (0)	100 (67)

NOTES: Number of cases is given in parentheses.
Table 3-8 is based on the subset of the positions used to construct Table 3-6 for which the
gender of the person who accepted the position was known. We do not know from these
data whether the person who accepted the position is the same person who received the
first offer, even in those cases in which the gender is the same. Number of cases is given in
parentheses.
SOURCE: Survey of departments carried out by the Committee on Gender Differences in
Careers of Science, Engineering, and Mathematics Faculty.

We do not have information in our survey data to permit investigating this
difference further. One plausible explanation is that many women who are offered
positions are the only woman interviewed for that position. If the only woman
interviewed is offered the position and turns it down (for whatever reason), that
position will inevitably be filled by a man. In fact, only one woman was interviewed for 205 (38 percent) of the tenure-track and 23 (24 percent) of the tenured
openings for which more than one person was interviewed. While there are many
reasons why a person might turn down a job offer, in this particular instance, it
is possible women, who are interviewed at disproportionally higher rates, also
receive more offers than men and have to turn some of them down.

FACULTY PERSPECTIVE ON HIRING

Turning to the faculty survey, the committee asked faculty who were either tenure-track or tenured and had been hired after 1996 what were their "main considerations in deciding to work for their current institution." Respondents could check up to 15 choices (the 15th and final choice was Other). For each selection, respondents could check yes or no. These data were coded for analysis as follows: If a respondent selected yes or no for some choices but left others unchecked, the unchecked choices were recoded as no. A chi-square (χ^2) test was conducted on each of the 14 substantive selections against gender to investigate whether women and men weighed factors differently when deciding to accept an offer for a position. The responses are presented in Appendix 3-8 and are summarized in Figure 3-2 below. The effect of gender was statistically significant only in the case of family-related reasons. As might have been anticipated, women were more likely to weigh family-related factors more heavily than men when deciding whether to accept an offer, but the difference is not substantial.

INSTITUTIONAL POLICIES FOR INCREASING THE DIVERSITY OF APPLICANT POOLS

Our findings suggest that once women apply to a position at a research-intensive institution, the chances that they will be invited to an interview and be offered a position are disproportionately high for many of the disciplines we surveyed. Yet the proportion of women in faculty positions continues to be low despite increasing numbers of women receiving doctorates in the sciences and engineering. In this light, and given that the percentage of women applying for positions is apparently lower than the percentage of women receiving Ph.D.s in the six target disciplines, it appears that the only strategy to increase female representation in the faculty ranks is to increase the percentage of women in the applicant pool.

The NRC's *To Recruit and Advance: Women Students and Faculty in Science and Engineering* (2006) identified institutional characteristics, culture, and policies that may have an impact on the percentage of females who choose to apply to academic positions in science and engineering. Some of these include:

- Increased institutional efforts in signaling the importance of a gender-diverse faculty. This might be accomplished by increasing the frequency of positive declarative institutional statements, by establishing a committee on women, by exercising close oversight over the hiring process, or by devoting additional resources to hiring women.
- Modified and expanded faculty recruiting programs. Consider, for example, creating special faculty lines earmarked for female or minority candidates, ensuring search committees are diverse, encouraging inter-

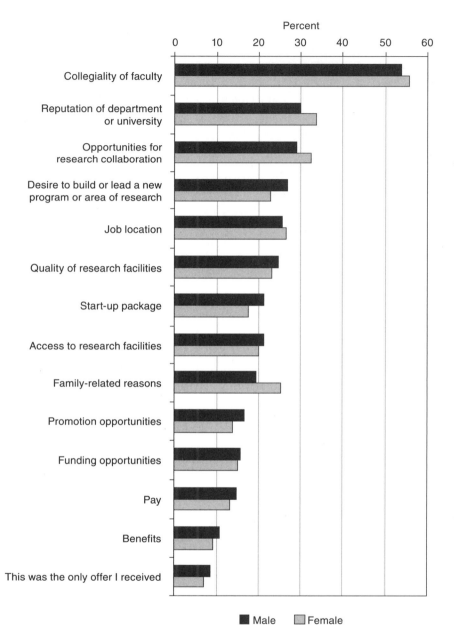

FIGURE 3-2 Main considerations for selecting current position (percent saying "yes, this was a factor"), by gender (see Appendix 3-8).
SOURCE: Faculty Survey carried out by the Committee on Gender Differences in the Careers of Science, Engineering, and Mathematics Faculty.

vention by deans when applicant or interview pools lack diversity, and
systematically assessing past hiring efforts.

- Improved institutional policies and practices. These might include insert-
  ing some flexibility into the tenure clock, providing child care facilities
  on campus, establishing policies for faculty leave for family or personal reasons, significantly stepping up efforts to accommodate dual
  career couples, and continuing to offer training at all levels to combat
  harassment and discrimination and to raise the awareness of all campus
  citizens.[11]
- Improved position of candidates through career advising, networking,
  and enhancing qualifications.

While all the strategies above might have an impact on the proportion of
women in applicant pools, it appears that only the last two might actually encourage more women to choose academia for their professional activity. The issue is
not whether female applicants are treated fairly in the interviewing and hiring
process; by several indications, they are. Where progress can still be made is in
attracting more women to academia by encouraging more of them to apply for
faculty positions at Research I institutions. It seems that refocusing resources to
develop strategies to encourage female graduate students to pursue a career in
academia has the potential for enormous impact.

Written policies and handbooks for faculty searches frequently note specific steps that can be taken to improve the diversity of applicant pools. These
include:

- Defining searches broadly to encourage a more diverse applicant pool;
- Posting the job advertisement in a wide range of outlets;
- Contacting professional associations that represent women (e.g., the
  Caucus for Women in Statistics, Society of Women Engineers, Associa-
  tion for Women in Science, etc.); and
- Evaluating the applicant pool during the search to determine if sufficient
  numbers of women are applying.

Departments reported a variety of actions in response to our survey question,
"What steps (if any) has your department or institution taken to increase the gender diversity of your candidate pool?" This was an open-ended question, and the
most frequent responses are shown in Table 3-9. Four hundred seventeen departments responded. Departments wrote in with answers ranging from zero to 6 steps
and citing anywhere from having zero to 15 policies in place. Targeted or special

---

[11] However, analysis presented in this chapter does not find an effect of the number of family-
friendly policies on the percentage of female applicants. The impact of such policies on applications
may bear further study.

TABLE 3-9 Steps Taken to Increase the Gender Diversity of the Candidate Pool

Step	Number of Departments Reporting
Targeted or special advertising	80
Other	71
General advertising	58
Recruiting at conferences, contacting women directly, using personal contacts	47
Help from diversity/EEO office or coordinator	47
Contacting colleagues and other universities	42
Special language used in advertising	34
Special consideration to females (e.g., making extra effort to interview females)	34
Informal networks	25
Grants or special funds for hiring women	19
Target of opportunity	19
Use of special databases or directories	18
Having a diverse search committee	17
Broadening searches	11

NOTE: Many of the 417 departments provided multiple answers to the open-ended survey question, and 71 departments that reported that they have taken steps other than those listed in the table.
SOURCE: Survey of departments carried out by the Committee on Gender Differences in Careers of Science, Engineering, and Mathematics Faculty.

advertising was the most frequently cited action, followed by general advertising. These were followed by recruiting at conferences, contacting women directly, and using personal contacts and assistance from on-campus diversity offices.

In addition, for most departments the total number of steps taken was not large. As shown in Table 3-10, 23 percent reported taking no specific action, and 43 percent reported taking just one. Only slightly more than 10 percent reported taking three or more steps.

SUMMARY OF FINDINGS

The analyses in this chapter reveal a number of important findings about the application, recruitment, interview, and hiring process.

TABLE 3-10 Number of Policy Steps Taken by Departments

Number of Departments	Number of Steps Reported Taken
96 (23)	0
178 (43)	1
98 (24)	2
34 (8)	3
10 (2)	4
0 (0)	5
1 (0)	6

NOTES: Numbers in parentheses are the percentage of all responding departments; 417 departments responded. Of these, 98 (24 percent) took two policy steps to increase the gender diversity of the candidate pool.
SOURCE: Survey of departments carried out by the Committee on Gender Differences in Careers of Science, Engineering, and Mathematics Faculty.

Applications

Finding 3-1: Women accounted for about 17 percent of applications for both tenure-track and tenured positions in the departments surveyed. There was wide variation by field and by department in the number and percentage of female applicants for faculty positions. In general, the higher the percentage of women in the Ph.D. pool, the higher the percentage of women applying for each position in that field, although the fields with lower percentages of women in the Ph.D. pool had a higher propensity for those women to apply. The percentage of applicant pools that included at least one woman was substantially higher than would be expected by chance. However, there were no female applicants (only men applied) for 32 (6 percent) of the available tenure-track positions and 16 (16.5 percent) of the tenured positions.

Finding 3-2: There are statistically significant differences in the percentage of women in the tenure-track and the tenured applicant pools across the six disciplines surveyed. Biology, chemistry, and mathematics had significantly higher percentages of female applicants than did all other disciplines. The percentage of female applicants in civil engineering, physics, and electrical engineering was significantly lower. The percentage of females among applicants to tenured positions was similar to the percentage of females among applicants to tenure-track positions.

Finding 3-3: In all six disciplines, the percentage of applications from women for tenure-track positions was lower than the percentage of Ph.D.s awarded to women. There were substantial differences among the disciplines. In civil engineering, electrical engineering, mathematics, and physics, the percentage of women applying for faculty positions was only modestly lower than the percentage of women receiving Ph.D.'s. However, in the fields with the largest representation of women with Ph.D.s—biology and chemistry—the percentage of Ph.D.s awarded to women exceeded the percentage of applications from women by a large amount (Table 3-2).

Finding 3-4: The median number of applications a department received for tenure-track jobs was 52 applications from men and 8 applications from women—or about 7 applications from men for every application from a woman. For tenured positions, the median number of applications a department received was 40 applications from men and 8 from women, for a ratio of 5 to 1. (Figure 3-1)

Finding 3-5: For job openings where only individuals of one gender applied, the gender was more likely to be male. There were no female applicants (only men applied) for 32 tenure-track positions or about 6 percent of available positions. Similar findings were seen for tenured positions. No women applied to 16 tenured jobs—or 16.5 percent of the positions. Most of the cases (29 of 32) when only men applied occurred in physics or the engineering fields.

Finding 3-6: Five factors were associated with the probability that at least one female would apply for a position, including (1) the type of position ($p <$ 0.0001); (2) the number of family-friendly policies in effect at the institution ($p = 0.001$); (3) a set of discipline indicators ($p = 0.03$); (4) prestige of the institution ($p = 0.04$); and (5) type of institution (approaches significance $p = 0.08$). No other factor was statistically associated with the probability of there being at least one female applicant.

Recruitment

Finding 3-7: Most institutional and departmental strategies for increasing the percentage of women in the applicant pool were not effective as they were not strong predictors of the percentage of women applying. The percentage of women on the search committee and whether a woman chaired the search, however, did have a significant effect on recruiting women. Most steps (such as targeted advertising and recruiting at conferences) were done in isolation, with almost two-thirds of the departments in our sample reporting that they took either no steps or only one step to increase the gender diversity of the applicant pool. (Tables 3-9 and 3-10)

Finding 3-8: The percentage of women on the search committee and whether a woman chaired the committee were both significantly and positively associated with the percentage of women in the applicant pool ($p = 0.01$ and $p = 0.02$, respectively).

Interviews

Finding 3-9: Across all the positions—tenure-track or tenured—an average of four men and one woman were interviewed for any particular position. Our survey data allowed us to examine the actual behavior of departments for the 545 tenure-track and 97 tenured openings for which we have gender data for applicants, interviewees, offers, and ultimate hires.

Finding 3-10: The percentage of women who were interviewed for tenure-track or tenured positions was higher than the percentage of women who applied. For each of the six disciplines in this study the mean percentage of females interviewed for tenure-track and tenured positions exceeded the mean percentage of female applicants. For example, the female applicant pool for tenure-track positions in electrical engineering was 11 percent, and the corresponding interview pool was 19 percent. (Table 3-3)

Finding 3-11: Although the percentage of women in interview pools across the six disciplines exceeded the percentage of women in applicant pools, no women were interviewed for 28 percent (155 positions) of the tenure-track and 42 percent (42 positions) of the tenured jobs. These figures are substantially higher than those for the men. However, the percentage of male applicants was much higher than the percentage of female applicants, and part of this number was comprised of cases for which there were no female applicants. In 23 percent of the tenure-track job openings (124 positions), at least 1 woman applied, yet no women were interviewed. In 25 percent of the tenured jobs (23 positions), at least 1 woman applied, but no women were interviewed. No men were interviewed for 3 percent (18 positions) of the tenure-track positions, and in one-half of those cases, there were no preceding male applicants; for 4 percent (4 positions) of tenured jobs, and in one-half of those cases, there were no preceding male applicants.

Finding 3-12: For tenure-track positions, the percentage of actual interview pools in which only men were interviewed (no women) was smaller than would have been expected based on applications and interviews for the positions surveyed for each of the six disciplines. For tenured positions, this was the case for three of the disciplines surveyed. Put another way, the percentage of actual interview pools in these disciplines including women was larger than would have been expected. For tenure-track positions, there were significant differences in electrical engineering (35 percent actual all-male interview pools compared to

56 percent probability of all-male pools) and mathematics (13 percent actual pools compared to 33 percent probable pools).

For tenured positions, there were significant differences, again, in electrical engineering (42 percent actual all-male interview pools compared to 62 percent probability of all-male pools); mathematics (39 percent actual compared to 44 percent probable); and physics (32 percent actual compared to 35 percent probable). This was not the case for the remaining disciplines, including biology (25 percent actual compared to 18 percent probable; civil engineering (46 percent actual compared to 35 percent probable); and chemistry, which had the greatest difference (50 percent actual compared to 24 percent probable). (Table 3-4)

Job Offers

Finding 3-13: For all disciplines the percentage of tenure-track women who received the first job offer was greater than the percentage in the interview pool. Women received the first offer in 29 percent of the tenure-track and 31 percent of the tenured positions surveyed. Tenure-track women in all these disciplines received a percentage of first offers that was greater than than their percentage in the interview pool. For example, women were 21 percent of the interview pool for tenure-track electrical engineering positions and received 32 percent of the first offers. This finding is also true for tenured positions, with the notable exception of biology, where the interview pool was 33 percent female and women received 22 percent of the first offers. (Tables 3-5 and 3-6)

Finding 3-14: In 95 percent of the tenure-track and 100 percent of the tenured positions where a man was the first choice for a position, a man was ultimately hired. In contrast, in cases where a woman was the first choice, a woman was ultimately hired in only 70 percent of the tenure-track and 77 percent of the tenured positions. When faculty were asked what factors they considered when selecting their current position, the effect of gender was statistically significantly for only one factor—"family-related reasons." (Figure 3-2; Tables 3-7 and 3-8)

As several of these findings suggest, many women fare well in the hiring process at research-intensive institutions. If women apply for positions at research-intensive institutions, they have better-than-expected chances of being interviewed and receiving offers compared to male job candidates. The likelihood of receiving an interview and ultimately an offer was particularly high, relative to application rates, in fields where women were less well represented, such as engineering and physics. These findings suggest that many departments at research-intensive institutions, both public and private, are making an effort to increase the numbers and percentages of female faculty in the sciences, engineering, and mathematics. At the same time, women continue to be underrepresented in the applicant pool relative to their representation among the pool of recent Ph.D.s.

The next chapter examines more fully the day-to-day lives of academics once they are hired, considering whether there are disparities by gender in the areas of faculty workload, institutional resources, and perceptions of departmental climate.

4

Professional Activities, Institutional Resources, Climate, and Outcomes

Once Ph.D.s have been hired into an academic position, it is natural to ask, what happens next? The milestones of an academic career are hiring, tenure, and promotion. In the context of these decisions, a primary question must be whether male and female faculty are treated similarly while they are employed. Is the day-to-day experience of being a faculty member similar for men and women?

Equitable treatment and opportunity are important for several reasons. First, how a faculty member is treated affects the ability of that faculty member to do the best research and teaching of which he or she is capable. This in turn affects subsequent decisions on the part of the university about salary, tenure, and promotion. It also affects subsequent decisions on the part of the faculty member about whether to entertain outside offers and whether to leave that university for a position elsewhere. Furthermore, the equitability with which a faculty member is treated can contribute powerfully to whether a faculty member feels he or she is a central part of the enterprise, as well as to the faculty member's sense of well-being and satisfaction with his or her professional life.

As noted in Chapter 1, there was anecdotal evidence that women do not fare as well as men professionally, but such differences can be subtle and hard to detect. The survey data presented in this report will provide information that is relevant to this perception and will help clarify the current status for women in the six disciplines surveyed at research-intensive (Research I or RI) institutions. According to one commentator:

> The study initiated at the Massachusetts Institute of Technology (MIT) several years ago by Nancy Hopkins has now been replicated at several other institutions, including Cal Tech. The reports have shown that women in science and

> engineering faculty are more likely to report that they feel marginalized and
> isolated at their institution, have less job satisfaction, have unequal lab space,
> unequal salary, unequal recognition through awards and prizes, unequal access to
> university resources, and unequal invitations to take on important administrative
> responsibilities, especially those that deal with the future of the department or
> the research unit. The fact that this study has been replicated at other institutions
> says that this is not an MIT specific problem. This is a generalized problem about
> the way women faculty at research-intensive universities experience their career
> environment. (Tilghman, 2004:9)[1]

This chapter examines variables that could contribute to a faculty member's
ability to excel at teaching and research. It asks about factors related to equitable
treatment of male and female faculty at research-intensive institutions in the six
disciplines surveyed, whether there are gender differences in salary, publications,
or the inclination to remain at that university, and whether differential treatment
accounts for any gender differences in salary, publications, or the inclination to
move on. The variables of primary interest to us fall into three categories: professional life, institutional resources, and climate. Under professional life, we include
how much of each of the following a faculty member does: the amount of research;
the amount of teaching, advising, supervising, and mentoring; and the amount
of service to the university or broader community. Under institutional resources
sometimes provided to support a faculty member's teaching and research, we
include start-up funds, summer salary, travel funds, reduced teaching loads,
laboratory space and equipment, and staff (postdocs, research assistants, clerical support). Under climate, we include variables that can contribute to a faculty
member's sense of engagement or marginalization within the department and the
institution, such as whether the faculty member is mentored by more experienced
colleagues, whether the faculty member is asked to contribute to important decisions in the department and the university, and whether a faculty member regularly
engages in conversation about research and teaching with his or her colleagues.

Three initial comments are necessary prior to proceeding with the assessment.
First, there are dozens of factors that together comprise a faculty member's job,
from the number of students she teaches, to whether she has the newest equipment in her lab, to whether she thinks her peers are collegial. One major benefit
that studies of hiring, tenure, and promotion have is that there is a dichotomous
end point that helps to focus attention. The study of professional activities,
institutional resources, climate, and outcomes lacks this. Therefore, anchoring
the analysis is somewhat more challenging. Second, the following analysis is
descriptive. Essentially, what is reported here about professional life, institutional
resources, and climate is the average response of male and female faculty to a

---

[1] Shirley Tilghman, 2004, "Ensuring the Future Participation of Women in Science, Mathematics,
and Engineering," in National Research Council, *The Markey Scholars Conference: Proceedings*,
Washington, DC: National Academies Press, pp 7-12.

series of questions about their work habits and environment. In the final section of this chapter, we look at how professional life, institutional resources, and climate contribute to important outcomes, such as research productivity and salary. In these analyses, we attempt to control for as many factors as we can that might contribute to the outcome, but it is likely that there are additional relevant variables about which we have no data. Without all relevant controls accounted for in the analysis, the results need to be taken as preliminary and as an impetus for further, more sophisticated research, rather than a definitive statement on the existence of disparities between male and female faculty. Finally, it should be noted that the analyses presented here provide an aggregated, often average, view. That view is not inconsistent with some women having very few resources and some women having quite a lot, nor does it negate the possibility that individual women (or men) are discriminated against in their access to resources. The deviation around average individual accounts of satisfaction or dissatisfaction can reflect a difficult reality, even when the averages among male and female faculty are the same.

The next three sections focus on professional activities, institutional resources, and climate issues. Professional activities include teaching, research, and service. Institutional resources cover a gamut of variables, including lab space, start-up packages, and research assistants. Climate focuses on such issues as mentoring and collegiality. Several of the above factors are further disaggregated into a variety of component elements. To study whether male and female faculty members reported different experiences on these dimensions and variables, we examine four types of information. First and foremost is our survey of faculty in six disciplines in RI institutions.[2] A second valuable resource is the U.S. Department of Education's National Survey of Postsecondary Faculty (NSOPF), undertaken in 2004 ("NSOPF:04").[3] That survey queried respondents regarding the fall 2003 term and thus occurred in a similar timeframe as the faculty survey. The other two information sources used throughout the chapter are individual research studies undertaken by scholars and gender equity reports completed by RI institutions.

After reviewing the three elements of day-to-day careers, we turn our attention to faculty outcomes. In the fourth section, we ask whether there are differences between male and female faculty in publication rates, grant funding, laboratory space (which is both an institutional resource and an outcome), nominations for honors and prizes, salary, outside offers, or the inclination to remain at the current institution, and which professional life qualities, institutional resources,

---

[2] Because we performed a large number of t-tests on our faculty survey data, we will only report as significant those results with $p < .05$ in order to protect ourselves from false positives. Results near $p < .05$ will be reported as approaching significance. For the regression analyses on our survey data, reported in the final outcomes section of this chapter, we will report any results with $p < .05$ as significant. The reader will want to note that there are some instances in which the differences are statistically significant, but the absolute differences are quite small.

[3] We also performed a large number of t-tests on the NSOPF:04 data, so we followed the rule for reporting significance in these data that is described in the previous footnote.

or climate variables contribute to differences in these outcome variables. This section draws on research done by individuals or as part of institutional studies to examine the issues of retention and job satisfaction, as our survey did not gather data on these variables.

PROFESSIONAL ACTIVITIES

In this section, we examine the three key areas of professional activities that characterize the day-to-day job of a faculty member: teaching, service, and research. Different departments weigh the value of these three activities differently, but in the Research I institutions, research is likely to be a primary concern. It is commonly believed that women spend more time teaching or performing service-related activities and less time on research than male faculty.

A note about time spent in professional activities is necessary. There are two ideas here: how many hours male and female faculty work and how they divide up the time they spend. Several studies have looked at the number of hours male and female faculty work and have found they tend to work long hours and similar numbers of hours. For example, a self-assessment conducted by the University of Pennsylvania found both men and women work nearly 60 hours per week. The NSOPF:04 found that full-time, professoriate faculty at Research I institutions in science and engineering (S&E) worked about 58 hours per week on average (58.5 for women and 58.1 for men).[4] Rather than ask faculty members how many hours they work, our survey asked respondents how they divide their time among research, teaching, and service. That is what we report here.

Research

It is often assumed that men spend a greater percentage of their time doing research than women. The percentage of time spent on research or scholarship was combined with percentage of time spent seeking funding in our survey data. Overall, men reported spending a slightly greater percentage of their time on research activities than women: 42.1 compared to 40.0 percent. This difference, while approaching significance, is quite small in absolute terms. Drawing on similar faculty from the NSOPF:04, there was no significant difference between men and women in the time spent on research activities: 43.2 compared to 39.7 percent.[5]

---

[4] Data was created using the Department of Education's Data Analysis System (DAS) available online at http://www.nces.ed.gov/dasol/. Gender was used as the row variable. The column variable was average total hours per week worked. Filters were only Research I institutions; full-time employed; with faculty status; assistant, associate, or full professors; with instructional duties for credit; and with principal fields of teaching as engineering, biological sciences, physical sciences, mathematics, and computer sciences.

[5] See previous footnote on how the DAS analysis was conducted.

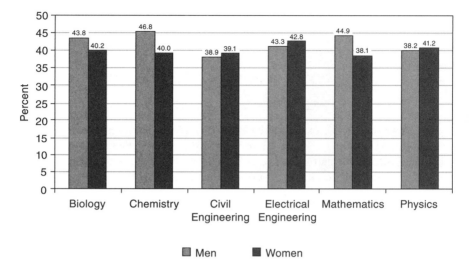

FIGURE 4-1 Mean percentage of time faculty spent on research (self-reported) by gender. SOURCE: Faculty survey carried out by the Committee on Gender Differences in the Careers of Science, Engineering, and Mathematics Faculty.

It is worth noting that the overall percentage of time faculty report spending on research activities is remarkably similar in the two surveys.[6]

Figure 4-1 shows the reported percentage time spent by faculty in research activities (including preparation of grant and contract proposals) disaggregated by gender and by discipline. Averages were computed over faculty who provided this information on the survey. To investigate whether there are differences in the percentage of time spent in research across disciplines or across genders, we fitted a simple linear model with percent time as the response variable and with discipline, gender, and the interaction between discipline and gender as the effects. We found no significant differences in percentage time spent in research, either across disciplines or between genders within discipline. Because comparing genders within discipline involved carrying out six comparisons, we used the Tukey-Kramer approach[7] to adjust the individual p-values. The smallest of the six p-values was obtained when comparing men and women faculty in chemistry (p-value = 0.217). All other p-values were above 0.35. Please note that discipline and gender accounted for a very small (about 2 percent) proportion of the variability observed in self-reported time spent in research activities. Thus, these p-values

[6] The committee acknowledges that the p-values for all the data presented for its faculty and departmental surveys are unadjusted and the fact that many of the data presented are interconnected.

[7] Kramer, C.Y., 1956, Extension of multiple range tests to group means with unequal numbers of replication, *Biometrics*, 12, 307-310.

are to be interpreted cautiously. A model in which other potential confounders are also included is presented later in this chapter.

In the NSOPF:04 data, there were no significant gender differences in any of the aggregated disciplinary groups reported (biology, physical sciences, mathematics, and computer science).

Teaching

In this section, the percentage of time spent on teaching, the number of classes taught, and the number of students advised are examined for gender differences. It is often assumed that female faculty spend a greater percentage of their time on instructional duties than male faculty.

Using the data from our faculty survey, the percentages of time men and women spent teaching and advising undergraduate and graduate students were combined and the average percentages were compared for men and women. Overall, female and male respondents reported spending approximately the same percentage of time on teaching and advising (men, 41.4 percent; women, 42.6 percent). The NSOPF:04 provided similar data: 44.2 percent for men and 42.0 percent for women. Here again, the percentages in the two surveys are remarkably similar.

Disaggregated by field, the difference between men and women faculty is approaching significance in chemistry and civil engineering, with women reporting more time spent on teaching and advising than men. In the NSOPF:04 data, there were no significant differences between men and women in the aggregated fields reported (biology, physical sciences, mathematics, and computer science).

Amount of Teaching

Our faculty survey also asked respondents how many undergraduate courses they were teaching in the current term/semester. In general, answers ranged from zero to two. There were no significant differences in the average number of undergraduate courses men and women were teaching (men, 0.83 courses; women, 0.82 courses; see Appendix 4-1). The NSOPF:04 data presented a similar picture, with a lower average number of undergraduate courses for women (men, 0.7 courses; women, 0.6 courses).

Looking at each of the six disciplines we surveyed, men were teaching marginally more undergraduate courses than women in electrical engineering; none of the other fields had significant differences between men and women. In the NSOPF:04 data, there were no significant gender differences in the teaching of undergraduate courses in the biological sciences, physical sciences, mathematics, and computer science. (There were too few cases to do this analysis for engineering faculty.)

The above analyses were repeated for graduate courses. Faculty teach fewer graduate courses, so here the distinction is between faculty who were doing no graduate teaching in the current term or semester and faculty who were doing some graduate teaching in the current term or semester. There was no significant difference found between men and women in terms of whether they were teaching graduate courses in our data (percent doing no graduate teaching: men, 50.8; women, 54.9; see Appendix 4-2.) or in the NSOPF:04 data (percent doing no graduate teaching: men, 46.8; women, 47.3).

There was no significant difference in any of the six fields we surveyed between men and women faculty in terms of whether they are teaching graduate courses. The data approaches significance in physics, where men are less likely to be teaching graduate courses than women. We conducted a similar analysis of the NSOPF:04 data and found that men were significantly more likely to be teaching graduate courses in the biological sciences (men, 65.8 percent; women, 59.7 percent) and in the physical sciences (men, 37.3 percent; women, 29.6 percent). In mathematics and computer science, there was no significant difference between men and women in terms of whether they taught graduate courses (men, 52.9 percent; women, 55.4 percent). (There were too few cases to conduct this analysis for engineering faculty.)

Finally, we explored whether gender is associated with the number of graduate thesis or honor thesis committees on which a faculty member serves. These data are shown in Appendix 4-6, and from the table, we see that the number of thesis committees on which faculty report serving is quite variable, ranging from zero all the way to 30. There appear to be some differences between men and women in terms of the numbers of committees on which they serve, but these differences appear to vary by discipline.[8] The NSOPF:04 asked faculty how many hours they spent on thesis and dissertation committees, and men spent marginally more time than women (men, 1.8 hours; women, 1.3 hours).

Service

There is a general awareness that female faculty spend a greater proportion of their time serving on departmental, school, or university-wide committees than men. In looking at the percentage of time faculty spend on service work, we combined the percentage of time spent on administration or committee work within the university with service outside the university. Overall, there was no difference between men and women in the percentage of time spent on service (men, 14.4 percent; women, 15.4 percent; see Appendix 4-7.). The NSOPF:04 found similar percentages of time spent on service, with no difference between men and women faculty (men, 16.1 percent; women, 14.8 percent).[9]

---

[8] The comparisons between men and women overall, and by discipline, in terms of the number of thesis committees a faculty member served on are not reliable, due both to small sample sizes and to the long-tailed distribution of this response; a few large values in response can strongly affect the comparison.

[9] Note that the definition the NSOPF uses is different from the definition used in the faculty survey.

Disaggregated by field, there appear to be no gender differences in the percentage of time spent on service in any of the six fields we surveyed (see Appendix 4-7). The NSOPF:04 found similar results (biology—men, 15.8 percent, women, 15.3 percent; physical sciences—men, 16.2 percent, women, 12.0 percent; and mathematics and computer science—men, 14.4 percent, women, 14.1 percent).

Committee Service

In addition to asking about the percentage of time spent on service, our faculty survey asked respondents how many committees they have served on. The view is that, in order to make committees more diverse, women are more frequently asked to serve on them, with the result that they serve on more committees than men do. The faculty survey asked respondents if they had participated in 10 types of departmental committees: undergraduate curriculum, graduate curriculum, executive, promotion and tenure, faculty search, fellowship, graduate admissions, facilities or space, program review, and "other." An initial variable was created that summed participation on the nine identified committees. While the actual range was between zero and nine, few faculty served on more than six committees, and disaggregated by field, there were many cells which contained no faculty members. Therefore, faculty members who served on six or more committees were aggregated into one category of those serving on six or more committees, so that at least one faculty member fit into each cell when the respondents were disaggregated by gender and field. Overall, the average number of committees served on was similar for men (1.61 committees) and women (1.76 committees) (see Appendix 4-8).

INSTITUTIONAL RESOURCES

This section focuses on a single, general question: do male and female faculty receive similar institutional resources? To explore this question, we examine a number of different resources. In order, they are start-up packages received on joining a department, summer salary, travel funds, reduced teaching loads, lab space, equipment, and support staff, including access to graduate research assistants (RAs) and postdocs.

Start-up Funds

Start-up packages are given to new faculty hires. A number of elements can be found in start-up packages, which makes it important to define clearly what is being quantified. Systematic surveys of start-up funds began in earnest around 2000. Examples include surveys conducted by the University of Colorado at Boulder in 1999 and surveys conducted by the Council of Colleges of Arts & Sciences—the New Hires Survey and the 2000 Big 10+ Chemical Engineering Chairs Survey. Summarizing their data, Ehrenberg and Rizzo (2004) write, "at research universities, these [start-up packages needed to attract new faculty members in the sciences] cost an average of $300,000 to $500,000 for assistant professors and often well over $1 million for senior faculty." A survey of start-up funds conducted by the Cornell Higher Education Research Institute (CHERI) in 2002 found:

> At the new assistant professor level, with few exceptions, Carnegie Research I universities provide larger start-up packages than other universities in the sample, and private research universities provide larger start-up packages than public universities. When the departments are broken down into four broad fields, physics/astronomy, biology, chemistry, and engineering, the average reported start-up package for new assistant professors at private Research I universities varied across fields between $337,000 and $475,000. Estimates of the average high-end (most expensive) assistant professor start-up package costs at these institutions varied across fields from $587,000 to $725,000.[10]

The data on start-up funds that is disaggregated by gender has been collected by individual institutions. A 2003 task force report at Princeton University, which collected data from five S&E departments, concluded "in the five departments examined, we found no statistical support for gender differences in start-up space, current space, or start-up financial packages. However, we did detect certain patterns. For example, the largest start-up packages have generally gone to men."

Both the committee's faculty survey and departmental survey requested data on start-up costs. On the faculty survey, faculty who were tenured or tenure-track and hired after 1996 were asked, "When you were first hired at this institution, how much were you given in start-up funds?" Respondents were asked to break down start-up costs into four categories: equipment, renovation of lab space, staff (e.g., postdocs), and other.

Summer Salary

The faculty questionnaire asked tenure-track or tenured faculty hired after 1996 whether they received summer salary funds when they were first hired at their current institution. Of those who responded, 71 percent of men and 68 percent of women indicated they did. When disaggregated by discipline, interesting differences appeared, with female faculty having a higher percentage in chemistry (81.8 percent compared to 71.2 percent for male faculty) who received summer

---

[10] The 2002 Cornell Higher Education Research Institute (CHERI) Survey on Start-up Costs and Laboratory Allocation Rules: Summary of the Findings is available at http://www.ilr.cornell.edu/cheri/surveys/2002surveyResults.html, accessed October 7, 2008. See also the presentation by Ronald G. Ehrenberg, Michael J. Rizzo, and George H. Jakubson, "Who Bears the Growing Cost of Science at Universities?" presented at the 2003 Conference. See also Ronald G. Ehrenberg, Michael J. Rizzo, and Scott S. Condie, "Start-up Costs in American Research Universities," CHERI working paper, WP-33, March 2003, Cornell University.

salary; while in mathematics the reverse was true, with 42.9 percent of male faculty as contrasted with 29.1 percent of female faculty (see Appendix 4-10).

Travel Funds

The faculty questionnaire asked tenure-track or tenured faculty hired after 1996 whether they received travel funds when they were first hired at their current institution. Of those who responded, 56 percent of men and 59 percent of women indicated that they did (see Appendix 4-11). Again, there was no substantial gender difference at this level of aggregation. There were some differences for men and women among the six disciplines in terms of the percentages of people receiving travel funds initially.

The survey also asked faculty respondents, "During the last five years, have you been given travel money by your department or institution to attend professional conferences or to conduct research offsite?" Of those who answered, approximately 42 percent of men and 43 percent of women answered yes.

Reduced Teaching Loads

Faculty may negotiate a reduced teaching load for an initial period after they are hired. New faculty often desire a reduced teaching load to allow them time to get settled in a new environment and to get their labs and their research set up and underway. The committee's survey asked all tenure-track and tenured faculty hired after 1996 whether they had received a reduced teaching load when hired. A large majority of new faculty reported receiving a reduced teaching load when they were hired (see Appendix 4-3). However, there was not a significant difference between men and women, in terms of the percentage who received a reduced teaching load when hired in any of the six fields surveyed.

Lab Space

Much of the discussion on lab space stems from the 1999 MIT report, *Report of the School of Science,* which found an "unequal distribution" of resources, including lab space, allocated to women.[11] This focused attention on the issue, and a number of other gender equity assessments at other universities have taken it up.[12]

Stanford's report, for example, found no disparity in lab space: "The Provost's Advisory Committee on the Status of Women Faculty on Thursday issued a variety

---

[11] Sara Rimer, "For Women in Science, Slow Progress in Academia," *New York Times*, April 15, 2005.

[12] See, for example, a thorough assessment conducted by New Mexico State University in 2003, "Space Allocation Survey," available at http://www.advance.nmsu.edu/Documents/PDF/ann-rpt-03.pdf.

of recommendations to strengthen the recruitment and retention of women faculty and, in a first-ever comprehensive analysis, has preliminarily found 'insignificant' differences between men and women in benefits and support such as laboratory space, equipment, start-up funds, research funds, and summer salaries" (James Robinson, *Report: 'No Pattern' of Disparity Between Men, Women Faculty,* Stanford Report, May 20, 2003).[13] The University of Pennsylvania found mixed results: "With respect to the professional status of women faculty, the committee determined that at the more junior ranks women had more research space per grant dollar than men, but women full professors averaged somewhat less space per grant dollar[14] than their male colleagues; in both SAS science departments and the School of Medicine, senior women faculty had about 85 percent of the space assigned to males."[15] Case Western Reserve found women had less lab space: "Despite these heavier workloads, participants believe that women often receive fewer benefits and support resources. Women tend to enter the university with more limited start-up packages. . . . They receive less space, have less access to graduate student assistance, and get fewer services from support staff."[16]

However, quantitative data on lab space are hard to find. It is critical that it be measured, because, as Purdue's report noted, it may be a perceptual or an actual discrepancy:

> Females responded differently than males on a number of these issues. However, most differences appear to simply reflect perceptual differences across the schools and the varying distribution of women in the schools (e.g., women are less satisfied than men with library resources, but this largely reflects the fact that Education and Liberal Arts schools, where faculty are the least satisfied with library resources, are also schools with relatively high proportions of women faculty).

> Taking into account these differences in gender representation across the schools, females are still less likely to believe that they have adequate laboratory space (48 percent) than are males (60 percent). In particular, women in agriculture, health sciences, and science are substantially less likely than their male counterparts to feel that they have adequate lab space.[17]

---

[13] Available at http://news-service.stanford.edu/news/2003/may21/womenfaculty-521.html.

[14] Note that the University of Pennsylvania's research used an unusual metric of research space per grant dollar.

[15] University of Pennsylvania Gender Equity Committee, "The Gender Equality Report, Executive Summary, Almanac, Vol. 48, No. 14, December 4, 2001, available at http://www.upenn.edu/almanac/v48/n14/GenderEquity.html. See the full report at: http://www.upenn.edu/almanac/v48pdf/011204/GenderEquity.pdf.

[16] CWRU Equity Study Committee, "Resource Equity at Case Western Reserve University: Results of Faculty Focus Groups," March 3, 2003, pp. 46-47. Available at http://www.case.edu/president/aaction/resourcequity2003.doc.

[17] Purdue conducted a survey in 2001, which asked female and male faculty whether they were satisfied with the amount of lab space. Women were less satisfied. (This is different from how much lab

The committee's survey asked faculty to identify how much lab space they have. It should be noted that lab space may mean different things to different people and in different disciplines. One problem, for example, is how to count shared lab space. Overall, lab square footage ranged from zero to 100,000 square feet. The two largest figures—47,000 and 100,000—both occurred in civil engineering and appear to be outliers.[18] Both observations were changed to missing. Estimated lab space was reported by 769 respondents. Overall, men reported significantly more lab space, with an average of about 1,550 square feet, than women, with an average of about 1,160 square feet. Disaggregated by field (see Figure 4-2), men had significantly more lab space in civil engineering and physics and marginally more in biology.[19]

One concern about studying lab space is that some faculty are theoretical while others are experimental, and the former might not need a lot of lab space. As the *Report of the Task Force on the Status of Women Faculty in the Natural Sciences and Engineering at Princeton* (2003:24) noted:

> Experimental science is heavily resource dependent. Consequently, in many departments, an individual's success is highly dependent on his/her access to space, equipment, supplies, students, postdoctoral fellows, and other laboratory personnel. At the time faculty members are hired, experimentalists are given laboratory space and start-up funds, which are used to purchase equipment and supplies as well as to support personnel. Funds for future support of the lab usually come from research grants, which are obtained from external, not University, funding. For faculty hired at the assistant professor level, additional laboratory space is usually needed to allow growth of research programs. Most experimentalists also require expensive equipment (e.g., electron microscopes, mass spectrometers, multi-node parallel processors) or services (animal care facilities, instrument specialists, technicians for common facilities and analytical labs) that are beyond the means of individual faculty members and that are purchased and/or maintained on a departmental basis.

This suggested a modified comparison conducted only on faculty who labeled themselves "experimental" or "both theoretical and experimental faculty" in the faculty survey. This comparison was conducted on 663 faculty and found that men who did experimental research reported a mean of 1,670 square feet of lab space, which was larger than the mean of 1,250 square feet reported by women who did experimental research. There are some interesting disciplinary differences. For example, in physics, men reported a median of 1,079 square feet of lab space and women report 800 square feet of lab space (see Appendix 4-12).

---

space each gender has.) Available at http://www.cyto.purdue.edu/facsurvey/faculty/survey/http://www. cyto.purdue.edu/facsurvey/faculty/survey/results/intro.htm.

[18] The medians for men and women faculty in civil engineering were quite similar, while the means were significantly different.

[19] Mathematics was dropped from this analysis, as only 11 respondents in mathematics reported having lab space.

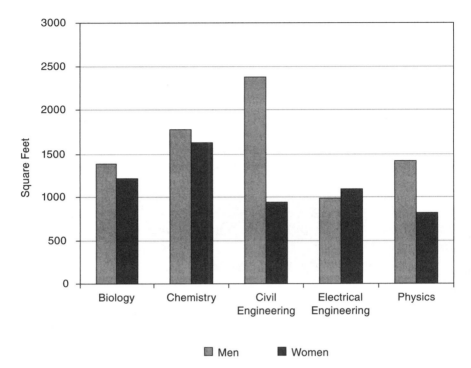

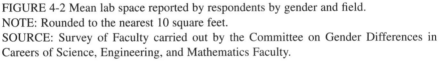

FIGURE 4-2 Mean lab space reported by respondents by gender and field.
NOTE: Rounded to the nearest 10 square feet.
SOURCE: Survey of Faculty carried out by the Committee on Gender Differences in Careers of Science, Engineering, and Mathematics Faculty.

The committee's survey also asked faculty who were hired after 1996 to report whether they had more, the same, or less lab space than they had when they were first hired. The analysis focused on comparing respondents who reported they had the same amount of lab space now compared to those who reported having more lab space now compared to when they were first hired. (Only 15 respondents noted that they had less lab space today.) A majority of both men and women reported no change in the size of their lab space (men, 72 percent; women, 70 percent). (See Appendix 4-13 for a multivariate treatment of this issue.)

Equipment

The survey asked respondents whether they had access to all the equipment they needed to perform their research. Three answers were coded: 2 = "Yes, I have everything I need," 1 = "I have most of what I need," and 0 = "I do not have access to major pieces of equipment that I need for my research." We dichoto-

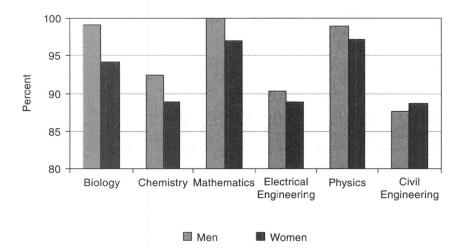

FIGURE 4-3 Percentage of men and women reporting having access to the equipment they need to conduct their research.
SOURCE: Survey of Faculty carried out by the Committee on Gender Differences in Careers of Science, Engineering, and Mathematics Faculty.

mized the answers to examine faculty who had access to an acceptable amount of the equipment they needed compared to those who did not (a "1" or "0"). Of those who responded, men were more likely than women to indicate that they had access to sufficient equipment (95 compared to 91 percent), which was a difference that is approaching significance (see Appendix 4-14). Disaggregated by field (see Figure 4-3 and Appendix 4-14), men were more likely to report that they had all the equipment they needed in chemistry and marginally more likely to report that they had all the equipment they needed in physics. We wanted to compare these results to data from the NSOPF:04. Unfortunately, that survey questionnaire does not include questions on satisfaction with equipment. The 1993 survey did ask respondents to rate the quality of "basic research equipment/instruments," "laboratory space and supplies," and "availability of research assistants," but those questions have been dropped from the more recent (1999, 2004) surveys.

Support Staff

The survey focused next on the number of research assistants (RAs) and postdocs supervised by the faculty, and the amount of available clerical support. For faculty, supervising RAs and postdocs is both an advantage and disadvantage. Such supervision may take a lot of faculty time and effort, yet support staff contributes a great deal to faculty research and publications, and thus, the availability of support staff can increase the productivity of faculty. Finally, supervising postdocs, who are the next generation of scholars, is recognized as one of the key activities of research faculty.

Faculty reported supervising between zero and 23 RAs, although 80 percent reported supervising between zero and 5 RAs. There was no difference between male (3.18 RAs supervised) and female (3.36 RAs supervised) faculty in the mean number of RAs supervised. Differences were small in every field.

Turning to postdocs, more than half of the faculty reported that they supervised no postdocs (see Appendix 4-15). The binary case (supervising no postdocs and supervising some postdocs) was then examined.[20] In general, there was no difference between male and female faculty in the probability of faculty who reported supervising one or more postdocs. The field with the largest difference was in biology where 49 percent of the women compared to 62 percent of the men supervised postdocs.

Finally, we examined clerical support. Faculty were asked whether they had all of the clerical support they needed, some of the clerical support they needed, or no clerical support. The variable was collapsed to focus on those who reported that they were satisfied compared to those who had less than they wanted or no access. In general, 54 percent of men reported that they had all of the clerical support that they needed, compared to 40 percent of women (see Appendix 4-16). Examined by discipline (Figure 4-4 and Appendix 4-16), men were more likely to report that they had all of the clerical support that they needed in chemistry and civil engineering.

CLIMATE

The Committee next examined some resources that may generally affect professional development. Here, the committee sought to assess whether male and female faculty were similarly engaged in their departments and institutions. There is a body of literature suggesting that women are isolated and marginalized. The former refers to not being part of the community in the department, institution, and more broadly (but not examined here), the scientific community. The latter refers to how much decision-making power women have on campus. In the wake of the 1999 MIT report, a number of universities conducted climate surveys on campus, discovering in some cases that female faculty face what is often termed a "chilly climate." For example, a 2003 climate survey of assistant, associate, and full professors by the University of California, Berkeley, found that women did not feel very included (Figure 4-5).

To examine isolation, our faculty survey collected data on mentoring, collaborative research, and interaction with colleagues. To examine marginalization,

---

[20] Specifically, the observation for any respondent reporting any non-zero number, in practice from 0.5 to 19 postdocs, was changed to 1.

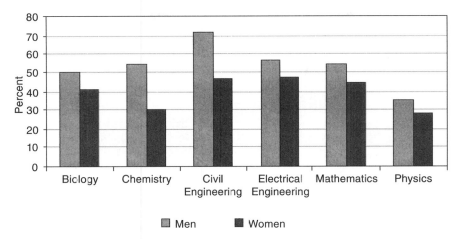

FIGURE 4-4 Percentage of faculty reporting having access to the clerical support they need, by gender and field.
SOURCE: Survey of Faculty carried out by the Committee on Gender Differences in Careers of Science, Engineering, and Mathematics Faculty.

we asked about participation in several types of committees (see Appendixes 4-4 through 4-8). We compared the number of women in the department with the number of women on search committees for hiring, on tenure and promotion committees, and engaged in other forms of university service.

Mentoring

Mentoring is often described as having significant positive effects on the retention and advancement of faculty. The survey asked tenure-track faculty and faculty tenured after 2001 whether they had or have a faculty mentor at their current institution. Among tenure-track faculty, 49 percent of the men and 57 percent of the women reported having a faculty mentor—a difference approaching significance. Among recently tenured faculty, 45 percent of the men and 51 percent of the women reported having a faculty mentor, which was not statistically significant. Disaggregated by field with tenure-track and recently tenured faculty combined (Figure 4-6), women were more likely to report having a mentor in electrical engineering and physics (see Appendix 4-17). Mentoring appears to be becoming more popular, and mentoring programs are spreading to more universities.[21] Thus, it is discouraging to find that only between half and two-thirds of

---

[21] See for example, Center for Research on Learning and Teaching (CRLT), The University of Michigan, "Resources on Faculty Mentoring." Available at http://www.crlt.umich.edu/publinks/facment.html.

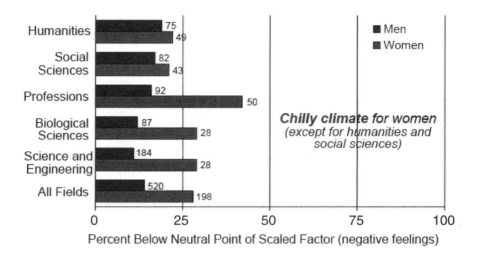

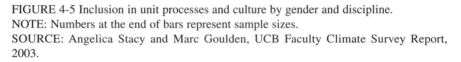

FIGURE 4-5 Inclusion in unit processes and culture by gender and discipline.
NOTE: Numbers at the end of bars represent sample sizes.
SOURCE: Angelica Stacy and Marc Goulden, UCB Faculty Climate Survey Report, 2003.

younger faculty have mentors in their home institutions. This does not seem to differ very much by discipline.

Collaborative Research

A second climate issue was whether the faculty member has been part of a research team at the institution. In this era of increasing collaboration and inter-disciplinarity on campuses, it was expected that most faculty would report that they had been part of a research team, and indeed, 65 percent of women and 62 percent of men responded that they had. By field, more than half of the faculty in every discipline except mathematics reported that they had been part of a research team, with no substantial gender differences.

Faculty Interaction

The survey also examined how often respondents interacted with each other. The survey asked, "Over the past year, how many faculty members did you discuss _____ with?" Respondents could answer: zero, one, two, three or more, or not applicable. The 10 topics were:

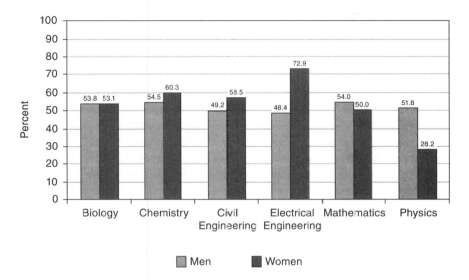

FIGURE 4-6 Percentage of faculty responding that they had a mentor by gender and field. SOURCE: Faculty Survey carried out by the Committee on Gender Differences in the Careers of Science, Engineering, and Mathematics Faculty.

Teaching
Research
Funding
Interaction with other faculty members
Interaction with administration
Climate in the department
Personal life
Family obligations
Salary
Benefits

We looked at whether there were significant gender differences for each issue separately. Men and women faculty did not differ in their reports of discussions with colleagues about 4 of the 10 issues (teaching, funding, interaction with administration, and personal life). Men reported significantly more discussion with colleagues about research, salary, and benefits than women. Men also reported marginally more conversation with colleagues about interaction with other faculty members and climate in the department. Only in the area of family obligations did women report marginally more conversations with colleagues than men reported.

Participation on Committees

Chairing committees was examined as one proxy for measuring marginalization—not having decision-making power within the department. As a first step, a variable was created to reflect the proportion of committees chaired by considering committees served on (i.e., the numerator is the number of committees chaired and the denominator is the number of committees served on, where the denominator is between zero and 9). Among the 1,063 faculty who served on at least one committee, 387 had chaired at least one committee. The variable was then dichotomized for faculty who participated on at least one committee into those who chaired at least one committee and those who chaired none of the committees on which they served. There was no significant difference between men and women in whether they chaired a committee on which they served (39 percent compared to 34 percent). An example of one of the committees reviewed is the chairing of undergraduate thesis committees. For this committee, there were disciplinary differences between male and female faculty in terms of chairing committees, with women chairing more committees than men in all fields except electrical engineering (see Appendix 4-5).

OUTCOMES

Professional activities, institutional resources, and climate can all be seen as inputs in the lives of faculty members. It is useful to know how faculty members spend their time, what resources are available to them, and how well integrated they are into the lives of their departments, universities, and disciplines. These factors are, however, only important to the extent that they contribute positively or negatively to a faculty member's ability to perform at the highest level in his or her teaching and research. Each of these factors may contribute directly to a faculty member's performance, or they may contribute to professional satisfaction and quality of life, which may in turn mediate between professional activities, institutional resources, and climate on the one hand and professional accomplishments on the other.

Teaching performance or effectiveness has been assessed in a variety of ways. The most prevalent is teaching evaluations (either by students or by peers), which are frequently used as performance indicators in salary, tenure, and promotion decisions (as well as for hiring decisions). Another possible approach that is interesting in principle but difficult in practice is the assessment of what students have learned in a course or while doing a project. One could also ask where a faculty member's graduate students land postdocs, faculty positions, or other employment. Unfortunately, this information was not gathered as part of our survey, and there are no national studies on these outcomes, with the exception of a number of studies on student evaluations of teaching. Some of those studies have looked at whether there are gender differences in the evaluations male and female faculty receive from students (Andersen and Miller, 1997; Centra and Gaubatz, 2000). However, a review of the literature suggests ambiguous results: "In many of these studies, male professors receive higher ratings than their female counterparts (Basow and Silberg, 1987; Kierstead et al., 1988; Sidanius and Crane, 1989). Others have female professors receiving higher evaluations than males (Tatro, 1995). Cashin's (1995) review of the literature showed little to no difference. Feldman's (1992, 1993) reviews found little to no difference in laboratory studies, while in observational studies, females had higher ratings in two-thirds of the cases" (Andersen and Miller, 1997:217). There could be a concern that women, who are particularly underrepresented in science and engineering, may be evaluated more harshly because they do not fit the perceived stereotypes of scientists or engineers.

In our survey, we asked about several important kinds of research performance or accomplishments. We asked respondents to tell us about their publications in the past 3 years, grant funding for their research, and how much lab space they had.

We also asked respondents to tell us about their salaries and whether they have been nominated for prizes or awards. Both salary and nominations for prizes or awards can be seen as indicators of the perceived quality of a faculty member's teaching and research. Finally, we asked our respondents whether they had received an outside offer in the past 5 years. This can be seen as an indicator largely based on the faculty member's research accomplishments.

We did not ask the respondents about their satisfaction with their professional lives. However, we do have data from the NSOPF:04, and we did ask our respondents whether they were thinking of leaving or retiring, which may be seen as an indirect measure of job satisfaction.

Research Productivity

Tenure and promotion often largely depend on research productivity, making it a crucial issue. Thus, the central question in this section: Is there a gender disparity in productivity? Faculty productivity can be defined in many ways. Here, we focus on grants received, lab space, and demonstrated, discrete output in the form of refereed publications and presentations.[22] Some of these can be the result of the efforts of a single faculty member or a collaboration between faculty members. For example, journal articles are sole authored as well as co-authored. Different disciplines place different amounts of emphasis on individual versus joint efforts. In the analysis of our survey data, we have combined individual and collaborative outcomes. In addition, one can measure productivity over the career of a scholar or more recently. For example, the NSOPF approach is to ask survey respondents to consider their scholarly activity both recently (over the past 2 years) and over

---

[22] Other possible measures include original discoveries and patents. On gender differences in patenting, see Ding et al. (2006).

their "entire career." Previous studies have found in the past that female faculty evidenced less research productivity than male faculty; however, this gap appears to be shrinking over time (Cole and Zuckerman, 1984; Fox, 1983, 1985; Long, 1992; NAS, NAE, and IOM, 2007; Xie and Shaumann, 2003). This suggests focusing on recent publications and grant funding, and on current lab space, which is what we have done in our survey data.

Publications

The survey asked respondents to report on the number of articles they had published in refereed journals and in refereed conference proceedings during the 3 years prior to the survey. Data for sole authorship and co-authorship were combined into a single variable.

We looked first at journal articles published in refereed journals. Overall, male faculty published marginally more journal articles in the past 3 years than female faculty (men, 8.9 articles; women, 7.4 articles). It is important to note that these statistics and those that follow related to publications could be misleading, given the significant interactions discovered in our multivariate analysis of gender, discipline, publications, and other variables. Disaggregated by field (see Figure 4-7), men appear to publish more papers than women in chemistry (men, 15.8; women, 9.4). The differences between men and women in mathematics and physics were smaller, with women publishing more than men in electrical engineering.

We then looked at the total number of publications in refereed journals and conference proceedings combined. Overall, there appears to be no difference between male and female faculty in the total number of publications (men, 13.9; women, 12.8). Disaggregated by field, men published significantly more than women in chemistry, but not in any other field; women published marginally more in electrical engineering, but not in any other field.

There are two differences between our survey data and the data from the NSOPF:04. The first is that we only asked about articles in refereed journals and in refereed conference proceedings, while the NSOPF:04 asked about articles both in refereed and nonrefereed journals, as well as books, textbooks, reports, and presentations. The second is that we asked respondents to sum information over the previous 3 years, while the NSOPF:04 asked about the past 2 years. A summary of the NSOPF:04 data is shown in Table 4-1. In these data, male faculty had significantly more publications than female faculty in the previous 2 years (men, 10.9 publications; women, 8.2 publications). Looking at the gender differences in the various subcategories, however, we find that the only significant difference between men and women was in articles in nonrefereed journals, a category we did not include in our survey.

Both the faculty survey and the NSOPF use simple numerical counts as measures of publications. Counts are one sensible approach, but they have their problems. "Simple counts of articles and books published account for neither

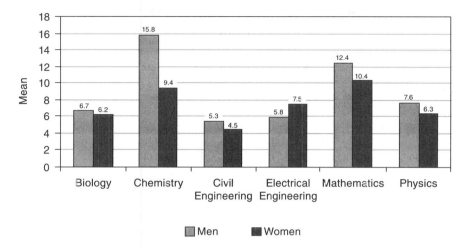

FIGURE 4-7 Mean number of sole or co-authored articles in refereed journals by gender and discipline.
SOURCE: Faculty Survey carried out by the Committee on Gender Differences in Careers of Science, Engineering, and Mathematics Faculty.

quality nor the importance of scholarship" (NSF, 2004b:8). Alternative approaches include weighting publications by the prestige of the source (e.g., top journals or university versus commercial presses for books) and counting citations of the publications to measure the impact of the faculty member's research. Both of these approaches are very difficult, but as the debate over quantity is increasingly clarified, taking an approach such as one of these may be fruitful in the future.

Next, we asked which variables might contribute to the number of articles a faculty member published in refereed journals and conference proceedings in the past 3 years. (Again, given the interactions, this more conditional look is more likely to accurately reflect the nature of the impact of gender and discipline on number of publications. Specifically, because disciplinary area interacts with gender and number of publications, one cannot directly interpret the effect of discipline in isolation from gender, and gender in isolation from discipline.) First we looked at the number of refereed journal publications. This model was fit to 1,404 observations corresponding to full-time faculty, tenured or tenure-track. Only 934 (of the 1,404 faculty) had complete information on all covariates in the model and had reported a number of journal articles. The number of journal publications is a count variable, making a Poisson model plausible for this outcome variable. We found, however, that a normal distribution was also a plausible model because the number of journal publications varied between zero and 40 with a mean of about 9. Therefore, to facilitate interpretation of results, we fitted an ordinary linear model to the number of journal publications and included the

TABLE 4-1 Average Measures of Recent Research Productivity by Gender

Measures of Recent Productivity	Mean for Men	Mean for Women
Total publications/scholarly works	10.9	8.2
Articles, refereed journals	7.2	5.9
Articles, nonrefereed journals	2.1	1.0
Books, textbooks, reports	0.6	0.4
Presentations	7.4	7.1

NOTE: Other possible measures include original discoveries and patents. On gender differences in patenting, see Ding et al. (2006).
SOURCE: NSOPF:04.

following covariates: gender, discipline, faculty rank, type of institution (public or private), prestige of institution, percent of time spent in research activities, having or not having a mentor, and all of the two-way interactions between gender and the other factors. The R^2 for the model was 19.0 percent (0.19).

Significant effects in the model were discipline (p-value < 0.0001), gender (p-value = .0001), rank (p-value < 0.0001), prestige (p-value = 0.0012), indicator for mentor (p-value = 0.005), percentage of time spent in research (p-value = 0.0001); and the three interactions gender with discipline (p-value = 0 .037), gender with rank (p-value = 0.042), and gender with mentor (p-value = 0.049).

Appendix 4-19 contains the least-squares (or marginal) mean number of journal publications for each level of each combination of fixed effects, and the lower and upper bounds of the 95 percent confidence intervals around each mean. Discipline has a very significant impact on the number of publications, as does gender, rank, prestige, and presence of a mentor. Also, discipline, rank, and presence of a mentor had modestly significant interactions with gender. Regarding the interaction of discipline and gender, we can assert that men publish more journal articles than women in biology; men publish more than women in chemistry; there is no significant difference between men and woman in mathematics, in electrical engineering, or in civil engineering; and men publish a borderline significant more than women in physics. Regarding the interaction of rank by gender, men increase the number of journal publications between the ranks of assistant and associated more than women do. The difference in the degree of increase from associate to full professor is less pronounced between the two genders. Regarding the interaction of mentor with gender, the difference between number of journal publications between men and women is more pronounced when faculty have mentors. Finally, the number of journal articles increases by 0.06 when a faculty member spends an additional 1 percent of his or her time in research activities.

The same analysis was then carried out in modeling total number of refereed publications (journal articles and refereed proceedings). Due to data quality issues, this analysis was conducted on 1,019 faculty members who reported both their refereed journal publications and their refereed proceedings articles, and whose values were not unrealistically high. Of the 1,019, only 774 had complete covariate information. The least-squares regression model with the same covariates obtained an R^2 of 23 percent. Since in this case no interactions were significant, it is easier to interpret the main effects. The significant effects were those of discipline (p-value < 0.00001), gender (p-value = 0.04), rank (p-value < 0.0001), prestige (p-value = 0.0002), mentor (p-value = 0.01), and percent time spent in research (p-value = 0.04). The marginal means and their 95 percent confidence intervals are provided in Appendix 4-19. Again, since none of the interactions among the main effects was significant, all the effects in this appendix can be interpreted in a straightforward manner. For instance, electrical engineers publish the most, followed by chemists, physicists, and civil engineers. Also, men publish more than women, and full professors publish more than associate professors, who in turn publish more than assistant professors. Furthermore, those at prestigious institutions publish more than those at less prestigious ones, and having a mentor increases the number of publications. Finally, since the regression coefficient of percent research time on fitting total number of refereed publications was 0.045 (with a p-value of 0.04), when research time increases by 1 percent, one can estimate that that will be accompanied by an increase in total publications of .045 per year.

Grants

In the sciences and engineering, grant activity is an important demonstrator of research ability. There are a number of approaches to comparing male and female faculty as grant recipients. Basic measures are whether faculty have received any grants and the total dollar value of any grants received. More in-depth measures include whether the grantee is a principal investigator (PI) or co-principal investigator (Co-PI), the source of grants, and the type of grants.

NSOPF:04 asked respondents whether any of their scholarly activity was funded. One hundred percent of both male and female full-time, professoriate faculty in S&E at Research I institutions indicated that it was. Our faculty survey found a lower percentage of faculty who responded that they were receiving sponsored research grants. It asked respondents, "What is the total dollar amount of the research grants on which you served as principal investigator or co-principal investigator during the 2004-2005 academic year? Include only direct costs for academic year 2004-2005." There were 213 faculty out of a total of 1,404 full-time faculty who did not provide an answer to the question. Of the 1,191 respondents, 163 (14 percent) answered zero, which means 86 percent of respondents (and 73 percent of all full-time faculty) reported some grant funding. This lower percentage found in our faculty survey compared to the NSOPF:04 is probably due to the fact that we asked faculty to limit their response to grants on which they served as a PI or Co-PI. Further, not all faculty may have interpreted the question in the same way; it may be that some among those reporting no funding had research funding during 2004 to 2005 but did not receive any *new* funding. A similar proportion of men and women (16 percent and 14 percent, respectively) had missing grant information. Thus, we base the remainder of the discussion on those faculty members who reported either no grant funding or some grant funding in the survey.

Women were more likely than men to report that they had grant income (88.6 percent compared to 83.8 percent). The University of Pennsylvania gender equity report found that women and men were equally likely to obtain grant support. As shown in Figure 4-8, disaggregated by field, women and men were equally likely to have at least one research grant on which they served as a PI or Co-PI, except in civil engineering (women, 99 percent; men, 88 percent) and mathematics (women, 77 percent; men, 62 percent), in which the differences were significant.

However, there may be differences when grants are examined in more depth. Although the University of Pennsylvania gender equity report found that men and women were equally likely to obtain grant support, it also found that men were more likely to be PIs, which suggests that an important focus for research would be which faculty have been PIs and which have been Co-PIs. Second, men and women may not be receiving the same types of grants. For example, while women's participation in National Institutes of Health grants is growing, the percentage is still quite small, and in some categories, the size of awards of the same type are smaller for women than for men (OER, 2005). However, a recent study by RAND (Hosek et al., 2005) found no gender differences in the amount of funding requested or awarded during the period 2001 to 2003 at National Science Foundation (NSF) and the U.S. Department of Agriculture (USDA). Interestingly, the study also found "gender differences in the fraction of first-year applicants who submit another proposal in the following two years" (p. xii). Women were less likely to reapply.

The next question in the faculty survey focused on the size of the grants received by the faculty member. About 6 percent of respondents answered that they had received $1 million or more in grant funding. The median response was $160,000.[23]

To explore the association between individual and institution-level factors and the success with which faculty raise research funding, we proceeded in two steps. We first modeled the binary outcome "grants/no grants" using logistic regression as a function of the following covariates: (1) gender, (2) disciplinary area (e.g., biology), (3) faculty rank (assistant, associate, or full), (4) type of institution (e.g., public versus private), (5) prestige of the institution, (6) number of publications,

---

[23] In addition, the following changes to the data were made: There were about a dozen observations in which respondents reported numbers of less than $1,000. It was assumed that these numbers, such as $60 or $100 actually meant $60,000 or $100,000. It was also assumed that a single entry of $1.3 was in fact $1.3 million.

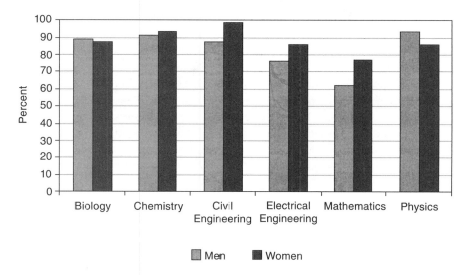

FIGURE 4-8 Percentage of faculty reporting having at least one research grant on which they served as a PI or Co-PI by gender and field.
SOURCE: Faculty Survey carried out by the Committee on Gender Differences.

(7) percent of time spent on research, (8) whether the faculty member reported that he or she had a mentor or not, and (9) all two-way interactions between the above covariates (see Appendix 4-20a and 4-20b for the analysis). This provided an estimate of the chance that a faculty member would or would not receive a research grant, regardless of the size. It is important to mention that whether a faculty member was or was not awarded a grant is of interest in itself because in some disciplines, receiving funding from a competitive agency is at least as important as the actual amount of funding received.

In a second modeling step, we estimated the amount of funding received *conditional on having at least some research funding.* The dependent variable was not the amount of grant funding, but instead the logarithm of the amount of grant funding to provide a dependent variable with a less skewed distribution, which can be useful in such models. There were 799 faculty (out of 1,191 who responded to the question about grants) with complete information for all model covariates. Of these, 697 (87 percent) reported receiving some grant funding during the period of interest.

We cannot conclude whether gender is associated with the probability of having a grant because the interaction between gender and discipline and between gender and rank were both statistically significant ($p < 0.05$). Women were significantly more likely ($p < 0.05$) to report having some grant funding in mathematics and in civil engineering, but the differences between the two genders were not significant in the other disciplines. Female full professors were significantly more likely to report some grant funding than their colleagues at the associate and assistant professor ranks, but for men, the gender rank interaction term was significant (see Appendix 4-20a and 4-20b). Overall discipline was significantly associated with the probability of having a grant, with faculty in electrical engineering being less likely ($p < 0.05$) to have a grant than faculty in the other disciplines. Faculty in civil engineering were significantly more likely ($p < 0.001$) than faculty in other disciplines to report some grant funding. The effect of discipline, however, is impossible to isolate, since the interaction between discipline and gender is highly significant, even after accounting for confounders such as rank, type of institution, and others. Assistant professors were less likely than associate professors ($p = 0.007$) and full professors ($p < 0.0001$) to have a grant, but there was no significant difference between full and associate professors. Again, discussing the effect of rank independently of gender is not reasonable, given the significant interaction between the two factors. Faculty at private institutions were equally likely to have a grant than those at public institutions ($p > 0.05$), and faculty at institutions of lower prestige were only marginally less likely ($p = 0.06$) to have a grant than faculty at institutions of either medium or higher prestige (which did not differ from one another). The number of publications a faculty member had was not associated ($p = 0.9$) with the probability of having a grant. Faculty who spent a greater percentage of their time on research were more likely to have a grant ($p < 0.01$). Finally, a faculty member who had a mentor appeared at first glance to be less likely to have a grant than a faculty member who did not have a mentor ($p < 0.0001$). However, this effect is difficult to interpret because the beneficial effect of the mentor depended on the gender of the faculty being mentored ($p < 0.03$). In fact, and contrary to what we might have anticipated, survey results suggest that the effect of having a mentor is not statistically significant among men. However, among women, a strong association between having a mentor and having grant funding was demonstrated.

Regarding the data on the interplay of gender and the availability of a mentor, we find in Table 4-2 that female assistant professors who do not have a mentor have a substantially lower probability of having a grant than female assistant professors who do have a mentor.[24] For male assistant professors, the presence of a mentor seems to make little difference. For associate professors, the presence of a mentor is associated with an increase in the probability of receiving a grant for both men and women, but the effect is much less pronounced than for female assistant professors.

We now consider the size of the grant and model the amount of funding as

---

[24] Inspection of the data revealed that the survey results were highly influenced by a single senior female faculty in civil engineering who reported having no grant funding, and she was removed from the survey results.

TABLE 4-2 Percentage of People Who Received Grant Funding by Gender, Rank of Faculty, and Mentor Status

Gender	Assistant	Associate
Males with mentor	.83	.93
Males with no mentor	.86	.86
Females with mentor	.93	.98
Females with no mentor	.68	.87

SOURCE: Survey of faculty carried out by the Committee on Gender Differences in Careers of Science, Engineering, and Mathematics Faculty.

a function of the same covariates. We used a log transformation on the response variable to better meet the normality assumption in the linear model. Because the log transformation collapses at zero, we added a negligibly small amount ($10) to the funding reports of zero. Figure 4-9 shows the distribution of funding amounts for all disciplines (except mathematics) in the log scale. Note that the distribution has a point mass at 2.3, which corresponds to the log of 10—the small amount added to grants of zero.

To explore the association between gender and other covariates on size of the grant, we considered all observations but fitted a Tobit regression model where the truncation bound was set to 3.0 (since the log of 10 is 2.30). That is, we obtained estimates of regression coefficients in the model that are unbiased and consistent once we account for the truncation. Of the observations, 485 exceeded the lower truncation bound and 221 did not.

There was no difference in the amount of grant funding received by male and female faculty after accounting for possible confounders of discipline, rank, type of institution, prestige of the institution, and research productivity (as measured by the number of publications).

Faculty in mathematics received grants of significantly smaller size ($p < 0.001$) than faculty in all other disciplines. Mathematics aside, the differences among all other disciplines were not statistically significant, with the exception of biology. Full professors had significantly larger grants than associate professors ($p < 0.0001$), who in turn had significantly larger grants than assistant professors ($p < 0.003$). Faculty at universities of highest prestige had significantly more grant funding (p-value = 0.002) than faculty at institutions of lower prestige, but the difference between the size of the grants received by faculty at the highest and the medium-prestige institutions was not significantly different ($p = 0.12$). Faculty at institutions of medium prestige had, in turn, marginally more funding than faculty at institutions of lower prestige ($p = 0.07$). These results are supported by the following average grant sizes: The average size of a grant obtained by faculty in 2004-2005 was $336,257, $352,639 and $463,231, respectively, at institutions of lowest, medium, and highest prestige. These values must be interpreted with caution. Making sweeping inferences

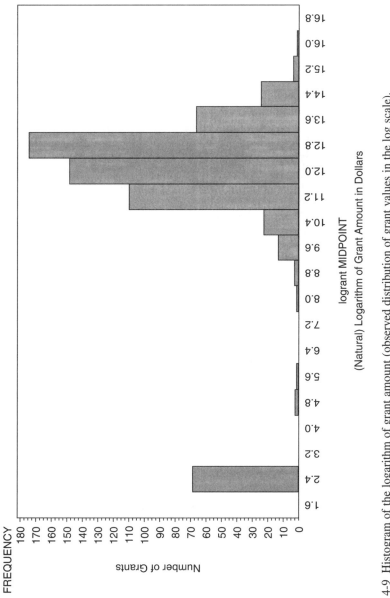

FIGURE 4-9 Histogram of the logarithm of grant amount (observed distribution of grant values in the log scale).

about funding levels across institutions, ranks, or even gender is unwarranted, given that we are considering only 1 year of funding data. Neither the number of publications nor the percent of time spent on research activities were associated with the size of grants obtained by faculty ($p = 0.3$ and $p = 0.7$, respectively); faculty with a mentor had less funding than faculty without a mentor ($p = 0.01$), which may reflect the fact that mentors are more prevalent among younger faculty who in turn tend to receive the smaller grants.

It is also of interest to investigate the association between gender and covariates *conditional on funding*. That is, if we were to consider only those faculty members who reported receiving some funding during 2004 to 2005, would results differ from those obtained when analyzing the entire set of outcomes? We anticipated that gender would not be associated with the amount of funding received by a faculty member even in the conditional analysis, given that gender was not found to be a predictor of the probability of receiving a grant or of the amount of grant funding *unconditionally*. A multivariate normal regression model fitted to the log-transformed *positive* grant values leads to approximately the same conclusions as the analysis that considers both zero and positive values together. Gender was still not associated with the amount of funding received, given that at least some funding was received. Full professors received significantly more funding than associates, who in turn received more funding than assistant professors. As before, the prestige of the university was positively associated with the amount of funding; the higher the prestige, the higher the average size of grants, everything else being equal. As before, we found that discipline was significantly associated with grant size, but essentially all of this effect is due to the fact that faculty in mathematics received significantly smaller grants than those in the other five disciplines.

Laboratory Space

We have already considered laboratory space as an institutional resource, because it is so crucial to the ability of faculty to get their research done. In our survey, male faculty reported having significantly more lab space than female faculty. This holds true both when we consider all faculty taken together and when we consider only those faculty who do experimental research. Here we look at what factors might contribute to the amount of lab space that a faculty member has and to the gender disparity in lab space.

Figure 4-10 shows the distribution of lab space in the entire sample (except mathematics) in the log scale. When space was reported as "0," a negligible amount (10 square feet) was added to allow for the transformation. Note that after the log transformation, lab space has a distribution that is approximately symmetric. Thus, a linear model was fitted to the log of lab space.

Explanatory variables in the model were gender, discipline, faculty rank, type of institution (public or private), prestige of the institution, grant funding, publi-

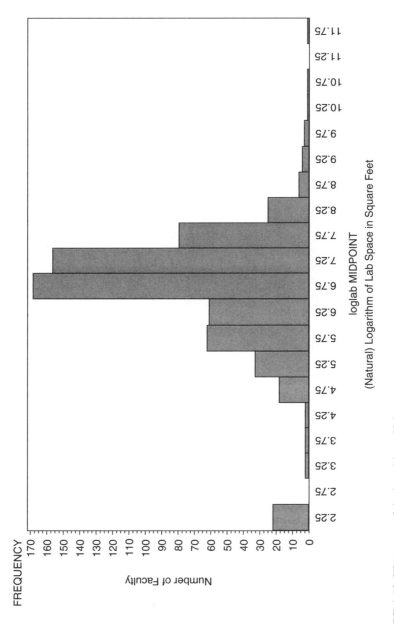

FIGURE 4-10 Histogram of the logarithm of lab space.

cations (refereed journal articles and conference proceedings), type of research (experimental, theoretical, both, educational, other), academic age (defined as time elapsed between receipt of Ph.D. and December 2004), and all two-way interactions with gender. Gender, discipline, rank, institution type, and prestige were classification variables; other variables were included as continuous. A random effect for institution was included, but the institution variance component was negligibly small. Because none of the interactions between gender and the other covariates were significantly different from zero, the model was re-fitted, but included only main effects. The model fitted the data reasonably well (R^2 = .32).

Significant associations with lab space were found for discipline ($p < 0.0001$), rank ($p < 0.0001$), type of institution ($p < 0.05$), prestige ($p < 0.01$), grant funding ($p < 0.0001$), research type ($p < 0.0001$), and publications ($p = 0.012$). Importantly, gender was not associated with lab space. Since, overall, when other variables were not taken into account, male faculty had significantly more lab space than female faculty, the absence of a significant gender difference in this analysis suggests that the overall difference is a function of gender differences in one or more of the other variables in the analysis. The most likely candidates were discipline and rank. We know that the percentage of female faculty varies between disciplines and between ranks. Therefore, if the disciplines and ranks with more male faculty were also the disciplines and ranks with more lab space, a simple comparison of the lab space of male and female faculty would show an overall advantage for men.

There are several interesting effects of variables on lab space. First, there was a positive association between grants and lab space. Everything else being equal, a faculty member who doubles his or her funding in a year can expect an 11 percent increase in lab space. Therefore, the effect of increased funding on space depends on the level of funding. A faculty member who has $10,000 in sponsored funding would only need to raise about $20,000 to increase his or her space by 11 percent. Yet someone who already has $100,000 would need to reach $200,000 in funding to have the same effect on his or her space. Second, not surprisingly, experimental researchers (most faculty call themselves experimental, and therefore research type was dichotomized to experimental or nonexperimental—and most women declared themselves to be experimental) reported having more lab space. Third, faculty at public institutions received more lab space than faculty at private institutions. Fourth, faculty at the most prestigious institutions reported having more lab space than faculty at institutions of medium prestige, who in turn report having more lab space than faculty at the least prestigious institutions. Fifth, our study indicated that the more senior faculty (those who have moved up in rank) had more lab space. This, however, is not a conclusion well supported by our data, which by their cross-sectional nature do not permit drawing longitudinal inferences. A snapshot impression can be misleading if, for example, senior faculty with large labs at a given point in time also had large labs when they were junior faculty. The effect of increase in rank and the effect of time itself are confounded when we can only explore faculty with a range of rank but during a single period of time. Finally, every additional publication, all else being equal, was associated with an increase in lab space of about 1 percent.

Nominations for Honors and Awards

Recognition in the field can be seen as another indicator of productivity, broadly defined, and as a goal that can bring greater job satisfaction, and perhaps indirectly affect such outcomes as likelihood of receiving a grant. One recent report examining the percentage of women nominated to an honorific society or for a prestigious award, and the percentage of women nominees elected or awarded from 1996 to 2005, found the percentages to be quite low (NAS, NAE, and IOM, 2007:128).

We asked respondents whether they had been nominated by their current department or institution for any international or national prizes or awards. Appendix 4-22 gives the number of faculty in each discipline who reported being nominated for at least one award at their current institution, as well as the number of missing responses in each discipline, by gender group. Overall, there was no gender difference in rate of nomination, with 28 percent of men and 26 percent of women reporting that they had been nominated. There were differences across gender when the data were disaggregated into the six disciplines we surveyed. Women were more likely to be nominated than men in electrical engineering and in civil engineering, and men were more likely to be nominated than women in biology and mathematics. Future research should also ask about nominations for university prizes or awards, and should ask separately about awards for research and those for teaching.

We looked at whether the probability that a faculty member would be nominated for an international or national prize or award was associated with various institutional or individual variables. There were 796 faculty with information for nominations and about all covariates in the model, and 240 of these reported having been nominated for an award. The probability that a faculty member would be nominated for an award was significantly associated with discipline, prestige of the institution, and type of institution. With one exception, none of the interactions between gender and any of the other variables was significantly related to the probability of being nominated for a prize or award. The exception was the interaction between gender and discipline, which was statistically significant ($p < 0.01$). This significant interaction prevents us from discussing the effect of discipline in isolation.

The probability that a faculty member would be nominated for an award was higher at private than at public institutions ($p = 0.03$). At institutions of high or medium prestige, faculty were either 1.5 or 5.5 times more likely, respectively, to be nominated for awards than at institutions of lower prestige. Not surprisingly, faculty with more refereed publications were more likely to be nominated for a prize or award than faculty with fewer, but this difference was not substantial.

Salary

It is fair to say that salary is an obligatory factor in every study that explores whether there are differences across gender in academic careers. Faculty salaries have been the subject of numerous university salary equity investigations, occasional lawsuits, and broader national studies. (See, for example, selected works by Barbezat, Becker, Bellas, Benjamin, Farber, Ferber, Ginther, Johnson, Perna, and Toutkoushian in the bibliography.) In general, studies suggest that women's salaries tend to lag behind men's. This, for example, is the conclusion that one would draw from the salary data that were collected by the American Association of University Professors (see Chapter 1). Data collected by the American Association for the Advancement of Science's salary and job survey (Holden, 2001, 2004) also support that claim. The gap, however, appears to be shrinking and our data, discussed below, confirm this tendency. Furthermore, while at first glance salary would appear to be a well-defined quantity that can be easily compared across gender, many factors appear to affect salaries in a complex way. Therefore, it is important to account for the potentially confounding effects of factors such as discipline, rank, productivity and others before attributing possible salary discrepancies to the effect of gender.

Here we examine salary information collected as part of our survey, as well as the salary data included in the NSOPF:04.

The faculty survey asked respondents to report their base salaries. We consider only the 1,404 full-time faculty who responded to the survey and who were assistant, associate, and full professors. There were 1,179 faculty for whom the salary information was not missing.[25] Appendix 4-21 shows the number of missing salary observations in each discipline and by gender. As is clear from the table, the proportion of faculty who did not respond to this question is similar across gender and across disciplines. The four observations correspond to two men and two women, and all exceeded $600,000 for a 9-month salary. The next-highest salaries reported were all below $250,000 for 9 months. One of the four outliers that were removed corresponded to a reported salary of almost $1.8 million, which is clearly unrealistic. About 20 percent of all respondents reported salaries below $100 for 9 months of work. Since these are likely to be values reported as thousands, we decided to multiply those reported salaries by 1,000 rather than lose the information. Two other salaries were removed from consideration and corresponded to two faculty members who, even after rescaling, ended up with 9-month salaries below $10,000 (the next lowest salary was $45,000). The wisdom of deleting the four highest salaries from the data set might be debatable, but from

---

[25] Out of the 1,179 respondents, 4 responses were considered to be outliers and were removed.

a purely statistical viewpoint is fully justified; the next-highest 9-month salary in the sample was $212,272, so the salaries at the high end of the distribution were clear outliers. Similarly, at the lower end, there was a clear gap between salaries below $10,000 and the next lowest at $45,000. Thus, for our initial salary analyses, we considered 1,173 faculty out of the 1,179 who responded.

Appendix 4-9 shows the mean salary by discipline, rank, and gender. It also shows the number of observations in each category. Statistics were computed using salaries standardized to a 9-month basis. Nor surprisingly, salary increases with seniority in all disciplines and both genders. Men appear to have a higher mean salary than women in almost all disciplines, but only among full professors. The difference between men and women seems to vanish for associate and assistant professors, and in some disciplines (e.g., electrical engineering and physics), female associate professors appear to receive a higher mean salary than their male colleagues. At the assistant professor level, the differences in mean salary are negligible and favor men or women, depending on the discipline. One interesting finding is that the *highest* salary among assistant professors is paid to a woman in *every* discipline, while the *lowest* salary is paid to women in only half the disciplines (mathematics, physics, and civil engineering).

In trying to understand the major predictors of salary, we first fitted a simple model that did not take into account potentially important factors, such as productivity. In this simple model, explanatory variables were gender, rank, academic age, discipline, and all the two-way interactions with gender. There were 1,169 observations with complete covariate information, and the model fitted the data well: $R^2 = 0.54$.

There were significant gender differences in salary in this model ($p = 0.009$), in which we controlled for several variables likely to differ between men and women (e.g., rank and discipline). However, the effect of gender cannot be interpreted in isolation of other factors because the interaction between gender and rank was also statistically significant ($p = 0.004$). Thus, we can only investigate whether salaries for men and women are similar within rank. We find that among full professors, men earn significantly more than women ($p < 0.05$). On average, male full professors earn about 8 percent more than female full professors. There are no significant differences in salaries for men and women among associate or assistant professors. Discipline and rank were also significant predictors of salary. Given these results, it seems likely that some of the gender differences in faculty salaries that are reported in other studies, in which rank and discipline are not controlled, are due more to the confounding factors rather than solely to any gender difference in salary.

We also looked at the salary data in the NSOPF:04. Because income is reported as a mean and there are small sample sizes in some disciplines, it was not possible to break out the analysis by field. In the NSOPF:04 data for full-time faculty who had instructional duties for credit and faculty status at an Research I institutions in engineering, biological sciences, physical sciences, mathematics, and computer sciences, we found that men received marginally larger salaries than women at the full professor level. However, there was no significant difference between the salaries of male and female faculty at the associate professor or assistant professor level. This assessment of gender differences in base salaries did not control for disciplinary differences or academic age, both of which are likely to have gender differences in them.

We were also interested in whether several additional variables predicted salary. To explore this, we again fitted a linear model to the log of base salary (9-month base), but now extended the list of explanatory variables in the model to include gender, discipline, faculty rank, type of institution (public or private), prestige of the institution, grant funding, publications (refereed journals and conference proceedings), academic age (defined as the time elapsed between award of the Ph.D. and December 2004), and all two-way interactions with gender. The model was fitted to the 753 observations with complete information for all covariates and resulted in an R^2 equal to 0.64. Gender, discipline, rank, institution type, and prestige were considered classification variables; the remaining were included as continuous variables. A random effect for institution was included, but the institution variance component was negligibly small.

Gender was *not* significantly associated with salary once other potential confounders were taken into account. Significant associations with salary were found for discipline ($p < 0.0001$), rank ($p < 0.0001$), prestige of the institution ($p < 0.0001$), type of institution ($p < 0.0001$), and grant funding ($p = 0.002$). The interaction between rank and gender was again significant ($p = 0.02$). Academic age and the interaction between grant funding and gender were approaching significance ($p = 0.07$ and $p = 0.09$, respectively).

As would be expected, full professors reported larger salaries than associate professors, who reported larger salaries than assistant professors. The more time that had elapsed since a faculty member received a Ph.D., the higher the salary, regardless of rank. The highest prestige institutions across all disciplines pay higher salaries than medium-prestige institutions, which in turn pay higher salaries than the lowest prestige institutions. Private institutions pay higher salaries than public institutions.

There was a positive association between grants and salary. Everything else being equal, a faculty member who increases his or her funding by a factor of 2.7 in a year would be linked with a 0.6 percent increase in salary. For a $75,000 annual salary, this would amount to an increase of $450, and so while this relationship is statistically significant, it has little practical significance. While the association between the number of publications and salary was not statistically significant, it was at least positive. Once again, please note that the discussion here somewhat overshoots the capabilities of our surveys. Because our surveys were cross-sectional, a conclusion stating that a faculty member who, over time, increases funding will also see an increase in her salary implies a longitudinal effect that our data cannot capture. Thus, these results, while plausible, must be cautiously interpreted.

Although women and men at a given rank appear to be compensated at similar levels (this does not apply for full professors), women may be at a disadvantage if they are less likely to be promoted to higher ranks. This topic is addressed in Chapter 5, where we found no evidence of differences among men and women in terms of promotion to higher ranks in our sample of full-time faculty at Research I institutions.

Outside Offers

Faculty retention and attrition focus on the likelihood that faculty will remain in a department. Some mobility is to be expected. Some faculty will move from one academic job to another or from academia to a position outside academia (e.g., in industry). Some faculty will leave departments to retire or because they are ill. The most problematic kind of attrition involves faculty who leave because they feel unwelcome. These faculty members have not failed, but they also have not fit in, and the departments they leave have invested time, money, and other resources that can be lost. For example, "new hires who leave their units in the first or second year end up costing programs tens of thousands of dollars in recruitment costs, moving expenses, start-up packages, and more" (Bugeja, 2004). The loss of a faculty member may also lead to a lost faculty line, as the faculty member might not be replaced.

It is often thought that female faculty attrition is greater than male faculty attrition, but the evidence is mixed (August, 2006; August and Waltman, 2004; Carter et al., 2003; Cohoon, et al., 2003; Trower and Chait, 2002; Yamagata, 2002). One way to examine retention issues, generally, is to ask faculty whether they have received outside offers or whether they are considering leaving their department.[26]

The survey asked tenured faculty whether they had "received an offer to leave their current institution in the last 5 years." Overall, the fraction of men and women reporting that they had received one or more offers was almost identical (32.7 percent of women and 32.5 percent of men.) There were no differences between men and women in biology and civil engineering (see Appendix 4-23). In chemistry and physics, a greater percentage of men than women reported they had received at least one offer to leave their current institution. This was reversed in mathematics and electrical engineering, where more women than men reported receiving one or more outside offers.

We looked at what factors contribute to getting outside offers. The response variable was considered to be dichotomous: zero, or one or more offers to leave the institution. The results of this analysis must be very carefully interpreted,

---

[26] Note that one reason to get an outside offer is to put pressure on a faculty member's current department to match a better offer. The faculty member might not actually want to leave his or her current department.

because clearly the people who received offers and stayed were the "happy" ones. Therefore, any statements regarding the effect of factors on offers to leave need to be qualified by noting that the findings are conditional on the fact that faculty remained at their institutions, whether they had offers or not.

In this analysis, we tried to investigate the effects of discipline, gender, rank, type of institution, prestige of the institution, whether the faculty member had a mentor, and the faculty member's productivity in terms of grant funding and publications on retention. However, we found that only 526 respondents (out of a total of 1,404 full-time assistant, associate, or full professors) had complete information for all covariates. Furthermore, only one male assistant professor and no female assistant professors reported receiving one or more offers to leave. Thus, we restricted attention to the 526 associate and full professors who had complete covariate information, and we fitted a model that included all the covariates of interest. The probability of receiving at least one outside offer was not different for men and women of any rank or across disciplines. In general, the probability of receiving one or more offers to leave was not associated with many of the covariates. The only two associations we found were with prestige of the institution and with research funding. As one might anticipate, faculty in institutions of medium or high prestige were more sought after than those at institutions of lower prestige ($p < 0.001$). Faculty with more research funding were also more likely to receive one or more outside offers. For every additional \$1,000 in research funding, the probability of having received at least one outside offer increased by about 1 percent.

Job Satisfaction

Job satisfaction is heavily intertwined with climate issues. Job satisfaction may be viewed as the expression of a faculty member's perception of engagement, power, treatment, and role, as well as departmental and institutional policies and procedures. It is a large and subjective area to tap into. It can also be a causal factor, affecting such outcomes as productivity and retention. Indeed, it seems likely that job satisfaction mediates, at least in part, between professional activities, institutional resources, and climate on the one hand, and the various outcome variables on the other. However, because we did not measure job satisfaction in our survey, we cannot test this possibility directly.

Satisfaction Data

Traditionally, most information on job satisfaction comes from surveys or focus group meetings undertaken by individual institutions or from the NSOPF. The results of many of these surveys suggest that women's job satisfaction falls below men's (Holden, 2001; Trower and Chait, 2002). Two recent national surveys have examined satisfaction with academic careers.

The NSOPF:04 asked several questions regarding satisfaction, although it did not probe very deeply into issues of workplace satisfaction. Four questions and responses from this survey are shown in Table 4-3. The mean responses show that female faculty are significantly less satisfied with their salaries and workloads than male faculty. Women are marginally less satisfied than men with their jobs overall. Men and women do not differ in their satisfaction with their benefits. This latter point is important, because some have interpreted the frequent finding that women are less satisfied with their jobs as indicating that women are generally more dissatisfied than men (or are more willing to express their dissatisfaction). The data in Table 4-3 show that women are less satisfied than men in particular areas rather than as a more general matter.

The Study of New Scholars, "Tenure-Track Faculty Job Satisfaction Survey (Trower and Bleak, 2004)"[27] examined full-time tenure-track faculty at six research universities. Important findings of the survey included "Females were significantly less satisfied than males with the following:

- Elements of work and expectations;
- Expectations for how to spend time;
- Expectations for research output;
- Expectations for the amount of outside funding needed;
- Time available for research;
- Resources available to support work; and
- Professional assistance for proposal writing and locating outside funds.

Relationships

- Commitment of the department chair to their success;
- Commitment of senior faculty to their success;
- Interest senior faculty take in their professional development;
- Opportunities to collaborate with senior faculty;
- Professional interactions they have with senior colleagues;
- Quality of mentoring they receive from senior faculty; and
- How well they fit in their department.

Diversity, Salary, Work–Life Balance

- Racial diversity of the faculty in their department;
- Ethnic diversity of the faculty in their department;
- Salary; and
- Balance between their personal and professional lives (p. 2)."

---

[27] The report, focusing on gender, is available at http://www.gseacademic.harvard.edu/~coache/downloads/SNS_report_gender.pdf.

TABLE 4-3 Satisfaction of Faculty with Employment by Gender

Satisfaction	Very Satisfied (%)	Somewhat Satisfied (%)	Somewhat Dissatisfied (%)	Very Dissatisfied (%)
Satisfaction with benefits				
Male	37.8	40.6	16.9	4.7
Female	38.6	42.6	12.7	6.1
Satisfaction with salary				
Male	25.4	43.4	22.1	9.1
Female	20.1	38.2	27.6	14.1
Satisfaction with workload				
Male	34.7	42.6	18.5	4.3
Female	23.2	50.8	20.2	5.8
Satisfaction with job overall				
Male	44.1	41.3	12.3	2.3
Female	39.1	45.3	13.4	2.2

SOURCE: National Center for Educational Statistics, 2004 National Survey of Postsecondary Faculty (NSOPF-04). Tabulation by NRC.

Individual universities have found similar results through surveys on their campuses. A 2002 survey of faculty at UCLA found the following: "Compared to male faculty, women feel less influential, rate their work environment as less collegial, view the evaluation process as less fair, feel less informed about academic advancement and resource negotiation, and rate the distribution of resources as less equitable."[28]

Dissatisfaction is an important concern. First, it is an obstacle to the success of faculty efforts in all areas of professional activities. It can have a negative effect on the collegiality and group decision making of a department. It may also be picked up on by undergraduate and graduate students, who may in turn feel discouraged about academic careers. While dissatisfaction may reflect problems in the workplace environment, it may also reflect pressures outside the workplace that affect women more than men and make it harder for them to get their work done.

Planning to Leave or Retire

Faculty who are planning to leave or retire, particularly the former, may be indicating that they are dissatisfied with their current work situation. Our survey asked faculty whether they were planning on leaving or retiring from their

---

[28] Gender Equity Committee on Academic Climate, 2003. *An Assessment of the Academic Climate for Faculty at UCLA,* Los Angeles, CA: University of California at Los Angeles.

current institution.[29] The variable was dichotomous; either the faculty member was not planning to leave or retire or they were. Out of 1,404 full-time faculty respondents, 171 did not provide an answer to this question, and the percentage of missing observations was essentially the same among men and women (13 percent compared to 11 percent, respectively). Both in general and disaggregated by field, there were no differences between men and women in their responses about whether they were planning to leave or retire (overall—men, 36.2 percent; women, 38.8 percent). Appendix 4-24 shows the percentage of men and women in each discipline who have indicated that they are not considering leaving or retiring from the institution. Since a larger percentage of men than women should be reaching retirement age, these data may suggest that more women than men are thinking of leaving their current institution for nonretirement reasons.

To get a clearer look at those faculty whose thoughts about leaving may more directly reflect dissatisfaction with their professional situation, we examined the percentage of faculty who were both planning to leave and had offers to leave in the past 5 years. We reasoned that faculty who were retiring were likely to respond positively to planning to leave but negatively to having recent offers to leave. Faculty who wished to change jobs were more likely to respond positively to both questions. There was no significant gender difference in the percentage of men and women overall who fit both conditions (men, 49 percent; women, 58 percent). Disaggregated by field, the difference between men and women was significant only in the case of electrical engineering, in which women were more likely than men to be considering leaving and to have received an outside offer.

SUMMARY OF FINDINGS

This chapter examined the day-to-day life of a full-time academic in S&E at Research I institutions. Principal findings can be found in four areas: professional activities, institutional resources, climate, and outcomes.

Professional Activities

Finding 4-1: There is little evidence overall that men and women spent different percentage of their time on teaching, research, and service. There is some indication that men spent a larger proportion of their time on research and fundraising than did women (42.1 percent for men compared to 40 percent for women). However, the difference only approaches significance, and the actual percentages of time that male and female faculty reported spending on research were not very different, with the exception of chemistry, for which men spent a signifi-

---

[29] In future studies, these two events should be separated, because male faculty tend to be older and are more likely to retire, while female faculty tend to be younger and are less likely to leave due to retirement.

cantly greater percentage of their time on research and fundraising (45.7 percent) than did women (39 percent) and mathematics (44.2 percent for men compared to 38.2 percent for women). (Figure 4-1)

Finding 4-2: **Male and female faculty appeared to have taught the same amount (41.4 percent for men compared to 42.6 percent for women).** There were no gender differences in the number of undergraduate or graduate courses men and women taught: 0.83 undergraduate courses for men compared to 0.82 undergraduate courses for women. The percentages not teaching graduate courses were 50.8 percent for men and 54.9 percent for women. (Appendix 4-2)

Institutional Resources

Finding 4-3: Men and women seem to have been treated equally when they were hired. The overall size of start-up packages and the specific resources of reduced initial teaching load, travel funds, and summer salary did not differ between male and female faculty.

Finding 4-4: Male and female faculty supervised about the same number of research assistants and postdocs. (Appendix 4-5)

Finding 4-5: There were some resources where male faculty appeared to have an advantage. These included the amount of laboratory space (considering both faculty overall and only those who do experimental research), access to equipment needed for research, and access to clerical support. At first glance, men seemed to have more lab space than women, but the difference disappeared once other factors such as discipline and faculty rank were accounted for. Insofar as the research a faculty member does is dependent on these resources, and the ability to accomplish as much as possible in turn determines his or her overall success, gender differences in these institutional resources could lead to gender differences in success.

At the same time, it should be noted that the apparent gender differences in access to these resources may reflect differences in access based on discipline or rank, because some disciplines and ranks have a higher perecntage of male faculty, and those disciplines and ranks could also have more lab space and equipment. This suggestion is supported by the finding that grant funding and research type (experimental versus nonexperimental) were significantly associated with the allocation of lab space. Since there are proportionately more male faculty than female faculty in some disciplines than in others, and since there are proportionately more male faculty than female faculty among full professors than among associate and assistant professors, it seems likely that the simple gender difference in lab space is actually a function of discipline and rank differences, as well as prestige of the institution. (Figure 4-2)

Climate

Finding 4-6: Female tenure-track and tenured faculty reported that they were more likely to have mentors than male faculty. In the case of tenure-track faculty, 57 percent of women had mentors compared to 49 percent of men. (Figure 4-6)

Finding 4-7: Female faculty reported that they were less likely to engage in conversation with their colleagues on a wide range of professional topics. These topics included research, salary, and benefits (and, to some extent, interaction with other faculty members and departmental climate). This distance may prevent women from accessing important information and may make them feel less included and more marginalized in their professional lives. Male and female faculty did not differ in their reports of discussions with colleagues on teaching, funding, interaction with administration, and personal life.

Finding 4-8: There were no differences between male and female faculty on two measures of inclusion: chairing committees (39 percent for men and 34 percent for women) and being part of a research team (62 percent for men and 65 percent for women). (Appendix 4-4 and 4-5)

Outcomes

Finding 4-9: Overall, male faculty had published marginally more refereed articles and papers in the past 3 years than female faculty, except in electrical engineering, where the reverse was true. Men had published significantly more papers than women in chemistry (men, 15.8; women, 9.4) and mathematics (men, 12.4; women, 10.4). In electrical engineering, women had published marginally more papers than men (women, 7.5; men, 5.8). The differences in the numbers of publications between men and women were not significant in biology, civil engineering, and physics. All of the other variables related to the number of published articles and papers (discipline, rank, prestige of institution, access to mentors, and time on research) show the same effects for male and female faculty. (Figure 4-7)

Finding 4-10: Although men were somewhat less likely to be a principal investigator or co-principal investigator on a grant proposal than were women, this difference disappeared when other variables were added in a regression analysis, where male and female faculty did not differ on the probability of having grant funding. Furthermore, because the effect of gender was confounded with the effect of rank and whether the person had a mentor, it is essentially impossible to isolate the effect of gender. The variables that appear to be associated with the probability of having a grant (discipline, faculty rank, being at a high- or medium-prestige university, and spending more time on research) had the same effect on male and female faculty. (Figure 4-8)

Finding 4-11: Male faculty had significantly more research funding than female faculty in biology; in the other disciplines, the differences between male and female faculty were not significant. There was no overall difference in the amount of grant funding received by male and female faculty, but there was a significant interaction between gender and discipline. The other variables related to the amount of grant funding (faculty rank, whether a faculty member is at a private university, whether a faculty member is at a university of higher prestige, having a mentor, and publishing more) were related in the same way for male and female faculty.

Finding 4-12: Female assistant professors who had a mentor had a higher probability of receiving grants than those who did not have a mentor. In chemistry, female assistant professors with mentors had a 95 percent probability of having grant funding compared to 77 percent for female assistant professors in chemistry without mentors. A similar but weaker pattern is exhibited for female associate professors. Over all six fields surveyed, female assistant professors with no mentors had a 68 percent probability of having grant funding compared to 93 percent of women with mentors. This contrasts with the pattern for male assistant professors; those with no mentor had an 86 percent probability of having grant funding compared to 83 percent for those with mentors. (Appendix 4-20a and 4-20b)

Finding 4-13: Overall, male and female faculty were equally likely to be nominated for international and national honors and awards, but the results varied significantly by discipline, making interpretation challenging. The other variables affecting the likelihood of being nominated for honors and awards (discipline, faculty rank, prestige of university, number of publications) affected this likelihood in the same way for male and female faculty. (Appendix 4-22)

Salary

Finding 4-14: Gender was a significant determinant of salary, but only among full professors. Male full professors made, on average, about 8 percent more than women, once we controlled for discipline. At the associate and assistant professor ranks, the differences in salaries of men and women disappeared. When we looked more broadly at variables that might predict salary, we found the following predicted salary in addition to discipline and rank: academic age (the amount of time between receipt of a Ph.D. and December 2004); prestige of the university; type of university (private versus public); and amount of grant funding. All these variables predicted salary in the same way for male and female faculty, with the exceptions of rank and grant funding, for which the beneficial effect of a grant was more pronounced for women.

Other Job Offers

Finding 4-15: Differences in the probability of receiving an outside offer for male and female faculty depended on discipline. In electrical engineering and in mathematics, women were more likely to have received an outside offer, while the trend was reversed in chemistry and physics. Men and women reported approximately the same probability of having received at least one outside offer in biology and civil engineering. The only two variables that predicted the likelihood of receiving an outside offer were prestige of the institution and the amount of grant funding, which demonstrated the same effect for male and female faculty. (Appendix 4-23)

Finding 4-16: There was no gender difference among faculty who were planning to leave and who had received an outside offer in the past 5 years,[30] except in electrical engineering, where women were more likely than men to be planning to leave and to have received a recent outside offer. The committee viewed these data as a measure of faculty dissatisfaction. (Appendix 4-24)

In this chapter, we set out to inquire if there were gender differences in the day-to-day academic lives of male and female faculty members. We wanted to determine if there were differences in professional activities, institutional resources, and climate, and if these influenced various important outcomes. Perhaps the most noteworthy finding is how much more similar the lives of male and female faculty seem to be based on our surveys, compared to the striking differences found in earlier research. The survey data indicate the importance of not simply relying on anecdotal information or past, individual experiences and emphasize the complexity of issues such as resource allocation and climate.

The overall data from this study send a positive signal about the institutional climates at Research I institutions, and this should encourage young women as well as men to pursue academic careers in math, science, and engineering, with the new awareness that their abilities rather than their gender will influence their experiences and their ultimate academic success.

Although the survey results do indicate that male and female faculty are encountering comparable opportunities in many ways, it is important to remember that these are group data. There may very well continue to be women who are experiencing fewer opportunities and less positive outcomes, at least in part because of their gender. Clearly, our survey questions, while extensive, were not exhaustive, and there were many areas not addressed.

---

[30] However, planning to leave or receiving outside offers are less than ideal proxies for job satisfaction. For example, faculty may plan to leave a position to retire.

Finally, this chapter focused only on the academic environment. There are other important factors not included in our survey that influence the participation and success of women faculty, particularly aspects of their personal lives such as family obligations. We hope that future research will be able to shed light on these critical areas.

5

Gender Differences in
Tenure and Promotion

National faculty data show that women continue to be underrepresented at the higher ranks of academia. While a partial explanation for the lower number of women at higher ranks (associate and full professor levels) has to do with the fact that women are newer entrants to academia—as noted in Chapter 3—a concern is that women faculty spend more time in lower ranks and are less likely to be tenured or promoted. In this chapter, we investigate whether women are, in fact, tenured or promoted at lower rates than men and find that national faculty data indicate otherwise. Controlling for the policies at their institutions, women who come up for tenure are tenured at greater rates than men, and women are promoted from associate to full professor at rates similar to those for men. The data, however, do not permit exploration of whether attrition prior to these career milestones occurs differentially by gender.

This chapter considers the advancement of women through the professorial ranks. It focuses on two critical junctures in most tenure-track faculty's careers: the awarding of tenure and promotion from assistant to associate professor, and promotion to full professor.[1] We do not discuss the transition of women from faculty positions to higher leadership positions (e.g., deans, provosts, or presidents) in academia.[2] To assess whether gender disparities might exist in the tenure and promotion process, the chapter examines three research questions:

---

[1] Some faculty remain associate professors and never come up for full professor status.

[2] For a discussion of issues and strategies related to bringing women into executive positions in academia, see NRC (2006).

- Are similar male and female faculty equally likely to receive tenure?
- Are similar male and female faculty equally likely to receive a promotion?
- Do male and female faculty spend equal amounts of time in professorial ranks?

Tenure and promotion decisions are designed to be based on merit. Although there may be some subjectivity in the determination of merit, the committee wished to compare rates of tenure and promotion for men and women who were similar along as many dimensions, such as experience and productivity, as could be observed. Assuming (1) men and women have similar talent, (2) are given similar amounts of time to demonstrate their teaching excellence, research potential, and commitment to service, and (3) are held to the same standard, then men and women should achieve similar tenure and promotion results. Different results would occur if one or more of these assumptions are false.

This chapter draws on evidence from the study's surveys of research-intensive (Research I or RI) institution departments in the sciences and engineering,[3] which the committee compared with data from other national surveys (primarily the National Science Foundation's Survey of Doctorate Recipients [SDR] or the National Survey of Postsecondary Faculty [NSOPF] of the National Center for Education Statistics at the U.S. Department of Education), as well as information drawn from gender equity studies carried out by individual institutions.

An important limitation of most analyses of tenure and promotion decisions is that they examine selected samples of those who succeeded in gaining tenure or promotion, and those who are eligible for these advances but have not yet been considered. Many studies examine the representation of women among tenured versus untenured faculty or among full versus associate professors. Generally, there are no data on the decision-making process itself.[4] One methodological approach is to make the argument that one would expect faculty who are 10 years beyond being hired as assistant professors to be tenured. It is then possible to compare the percentage of men and women who have in fact received tenure. This comparison, however, omits the faculty who left prior to being considered for tenure (possibly because they had been informed that they were unlikely to receive it), as well as those who were considered, but not awarded tenure. A second approach is to examine time spent in the assistant professor rank by those who were promoted to associate professor and time spent as associate professor by those who were promoted to full professor.

---

[3] The committee acknowledges that the *p*-values for all the data presented for the study's surveys of faculty and departments are unadjusted and that many of the data presented are interconnected.

[4] It may be that the only time the decision-making process becomes publicly visible is during litigation brought by faculty denied tenure or promotion.

Longitudinal data tracking individual academic career trajectories from first hire through tenure or departure are generally lacking, even in university-specific analyses. This is an important gap that can be readily remedied through the efforts of institutional researchers, with appropriate resources. This report's analysis of departmental data does, however, allow a direct examination of the tenure and promotion decisions since we asked departments to report on every tenure and promotion case they considered in the prior 2 years, independent of the outcome of the case. This allows us to obtain data on both successful and unsuccessful tenure cases. However, it will not overcome any bias due to attrition prior to these milestones.

The chapter first describes the nature of tenure and promotion processes in RI institutions. Second, it describes the outcomes of tenure and promotion decisions. Finally, the chapter uses multivariate methods to examine how tenure and promotion for men and women are affected by university programs and policies, such as changes to the tenure clock.

TENURE AND PROMOTION PROCESSES

Both tenure and promotion decisions are evaluations or reviews conducted by peers of a faculty member's professional activities, which lead to significant status changes. Tenure can be considered as a change from a probationary or fixed length appointment to an indefinite appointment. Such a change provides the faculty member with greater freedom in his or her professional activities or greater economic security, or both, although further promotions depend on continued research productivity and contributions to teaching and service.

Promotions are changes in status, such as from assistant to associate or from associate to full professor. The tenure decision and first promotion mark the transition between tenure-track assistant professor to associate professor. Most often tenure and promotion to associate professor occur at the same time, although some universities make these decisions separately. In the committee's survey, of 407 departments in RI institutions that responded, 318 (78 percent) granted tenure and promotion to associate professor together in a single decision. Disaggregated by discipline, Table 5-1 shows that 72 to 79 percent of departments decide tenure and promotion together, with the exception of chemistry, where 85 percent of 74 responding departments make one decision.

These decisions typically take place in the sixth year. Among the 407 responding departments, the modal response was 6 years elapsing between hiring and the tenure decision, with a range of 2 to 12 years. Fully 83 percent of departments indicated the tenure decision was made in the fifth or sixth year. As shown in Table 5-2, similar results were found within each discipline, with the exception of mathematics, where about one-quarter of departments responded that untenured faculty come up for tenure in 2 to 4 years.

TABLE 5-1 Percentage of Responding Departments That Decide Tenure and Promotion Together by Discipline

Discipline	Percent of Departments
Biology	75
Chemistry	85
Civil engineering	77
Electrical engineering	79
Mathematics	79
Physics	72

SOURCE: Survey of departments carried out by the Committee on Gender Differences in the Careers of Science, Engineering, and Mathematics Faculty.

From the vantage point of most tenure-track assistant professors, tenure may be the seminal event in their professional lives. Certainly, the first several years of academic life are spent building a dossier that will establish the case for granting tenure. The tenure decision grants substantial job security, validation of quality of work, possible monetary rewards (via salary adjustments), and increased institutional resources and authority. Universities typically have an "up or out" policy after a given number of years as a tenure-track assistant professor. If not granted tenure, the faculty member must leave his or her position for another position not on the tenure track or for employment outside the university. Faculty who believe they will not be granted tenure may choose to leave before facing the decision.

A second promotion marks the transition between associate professor and full professor; it is optional because some faculty may simply stay at the associate professor rank—although this was truer in the past than for current faculty. This decision also occurs several years after the first promotion.

These decisions are made, first, by a tenure or promotion committee comprised of departmental colleagues, typically followed by evaluation at successively higher administrative levels, such as a college- or school-wide committee, a dean, a provost or vice president for academic affairs, and the president of the institution. Exactly who is involved differs depending on the university, but such oversight is typical. The first decision would be made by the tenured departmental faculty, and administrators can support or reverse lower level decisions. Candidates generally provide a full curriculum vitae (C.V.); a statement describing their research accomplishments and goals, teaching history, teaching evaluations, and service to their department, university and profession; and copies of selected publications. Outside evaluations of research contributions of the applicant are generally solicited by the department from leading researchers located at other universities in the applicant's field. Sometimes internal evaluations are solicited from the same or other departments as the candidate's department.

TABLE 5-2 Number of Years Between Hiring and Tenure Decision

Number of Years

Discipline	1	2	3	4	5	6	7	8	9	10	11	12
Biology			1.37	1.37	20.55	65.75	6.85	1.37		1.37		
Chemistry				1.37	34.25	52.05	8.22	1.37	1.37		1.37	1.37
Civil engineering		1.85	1.85		37.04	51.85	5.56		1.85			
Electrical engineering					32.2	50.85	15.25		1.69			
Mathematics		1.43	12.86	11.43	27.14	44.29	2.86					
Physics			1.32	10.53	26.32	56.58	5.26					

NOTE: Numbers refer to percent of responding departments.
SOURCE: Survey of departments conducted by the Committee on Gender Differences in the Careers of Science, Engineering, and Mathematics Faculty.

Service on Tenure and Promotion Committees

Women are likely to serve on tenure committees but unlikely to chair them. As shown in Table 5-3, in 57 percent of tenure cases, there was at least one woman present on the tenure committee. In 690 cases for which the gender of the committee chair was known, it was a man in 90 percent of the cases. It was similar for promotion cases: in 51 percent of the promotion cases there was at least one woman present on the promotion committee. In 459 cases for which the gender of the committee chair was known, it was a woman in 13 percent of these cases.

EQUITY IN TENURE AND PROMOTION DECISIONS

The reason to ask whether there is equity in tenure and promotion decisions today is that there is a body of evidence suggesting that there is inequity. Specifically, the literature suggests that women as a group are less likely to receive tenure or a promotion (and it may take longer for women to reach those milestones). This section reviews several key studies on gender and tenure and promotion and identifies two reasons why there might be differences regarding rates of and time to tenure and promotion.

Several quantitative studies found that women were less likely than men to be tenured or promoted, or that women took longer to advance.[5] Examples include the National Research Council (NRC) (2001), Perna (2001a), Ginther (2001), and the National Science Foundation (NSF) (2004d). The NRC (2001) examined gender patterns in academic careers using data from selected years of the SDR: 1973, 1979, 1989, and 1995. Using a broad definition of Science and Engineering (S&E), which included the social sciences, and examining a wide range of higher education institutions, the NRC compared the percentage of men and women who had tenure among all tenure-track faculty. In 1995, 60 percent of women had tenure and 40 percent were tenure-track, while 79 percent of men had tenure and 21 percent were tenure-track. Second, the NRC examined men and women at different points in time in their careers, grouping men and women by the number of years that had elapsed since they received their Ph.D.[6] In examining men 1 year out, 2 years out, 3 years out, etc., the 2001 NRC report found a greater percentage of men were tenured than women (with the exception that a greater percentage of women were tenured among very recent Ph.D.s). Finally, using logit analysis, the NRC found that the difference between the percentage of men with tenure

---

[5] This general finding is commonly stated, even though individual institutions might have tenure or promotion rates that are comparable for men and women. As Nancy Hopkins (2006:18) notes in the case of the Massachusetts Institute of Technology (MIT), "Overall the tenure rates for men and women are almost identical in both the Schools of Science and Engineering." Looking at a broader segment of academia is thus necessary to see if MIT, to continue the example, is representative of many institutions or is an outlier.

[6] This is done by subtracting the year an individual received a Ph.D. from the survey year.

TABLE 5-3 Female Participation in Tenure and Promotion Committees

Field	Tenure Cases with at Least One Woman on the Committee (%)	Tenure Cases with a Woman Chair (%)	Promotion Cases with at Least One Woman on the Committee (%t)	Promotion Cases with a Woman Chair (%)
All fields	57 (768)	10 (690)	51 (508)	13 (459)
Biological sciences	74 (142)	11 (122)	71 (80)	20 (69)
Chemistry	55 (121)	11 (115)	58 (80)	7 (72)
Civil engineering	59 (115)	9 (108)	47 (73)	13 (63)
Electrical engineering	57 (113)	16 (102)	57 (95)	20 (89)
Mathematics	54 (147)	3 (125)	41 (86)	6 (77)
Physics	40 (130)	8 (118)	37 (94)	10 (89)

NOTE: Number of cases in each field is in parentheses.
SOURCE: Survey of departments carried out by the Committee on Gender Differences in Careers of Science, Engineering, and Mathematics Faculty.

and the percentage of women with tenure favored men, even when controlling for factors such as field, career age, and institution type. The 2001 NRC report also included individual factors such as citizenship, marital status, and family status, in addition to whether the institution was public or private. A parallel analysis for male and female full professors found similar results.

Perna (2001a) sought, among other questions, to assess whether the probability of being tenured or holding the rank of full professor was related to gender, after controlling for other factors that might affect the tenure and promotion decision. Perna used logit analysis on a different national data set, the 1993 NSOPF. Two findings are of interest: "Women and men who are participating in the tenure process appear to be equally likely to be tenured after taking into account other differences" (p. 561). On the other hand, the study notes, "Tenured women faculty at 4-year institutions are less likely than tenured men faculty to hold the highest rank of full professor even after controlling for differences in human capital, research productivity, and structural characteristics" (p. 561).

Ginther's analysis (2001) pooled cross-sectional samples of tenured or tenure-track faculty from the 1973 to 1997 SDR. She created a second analysis file by linking data on individuals who received a Ph.D. between 1972 and 1989 and who were sampled across several SDR waves. Ginther used probit models and duration models to assess whether there are gender disparities in the probability of "promotion to tenure." Her principal finding was "women are less likely to be promoted than men" (p. 20). Hazard analysis also suggested that women are about 12 percent less likely to be tenured.

Like the Ginther study, a recent study conducted by the NSF (2004d) used linked SDR data on individuals over time to examine whether gender was related to either particular outcomes on the career path or how long it takes "doctorate recipients to achieve career milestones" (p. 1). This study found that "women with eight or nine years of postdoctoral experience are about 5.9 percentage points less likely than men to be tenured. The comparable estimate for women with 14 or 15 years of experience is about 4.1 percentage points" (p. 3). Similarly, women were less likely to be full professors: "After accounting for controls, women with 14 or 15 years of postdoctoral experience who are employed full-time in academia are almost 14 percentage points less likely than men to be employed at the rank of full professor. The comparable estimate for women with 20 or 21 years of postdoctoral experience is similar" (p. 3).

Two competing hypotheses could underlie these findings. First, it could be that women present weaker cases for tenure due to lower productivity. Alternatively, women's lower rates of promotion could result from bias that causes women with equivalent qualifications to be judged less positively than similar male colleagues. With regard to the first hypothesis, the SDR provides some support for the case that female faculty produced less scholarly output in terms of numbers of publications. It has been proposed that women have fewer publications either because they receive fewer resources from their universities to support research, or because women spend less time on research. Although it is plausible that women could spend less time on professional activities if they are the primary caregivers at home and have more responsibilities outside of work, our data, presented in Chapter 4, show that in four of the six disciplines considered, women and men spend comparable percentages of their time in research-related activities.

The second rationale to explain why women might have a lower likelihood of receiving tenure or a promotion is evaluative bias on the part of their peers during tenure or promotion decisions. Bias may occur in several ways. First, women's research may be undervalued by colleagues. Second, women's teaching evaluations may not be as positive as those for men because of student bias. Third, women's external letters of recommendation may not be as positive.[7] However, determination of which two competing hypotheses provides the better explanation for why women take longer to achieve career milestones can only be addressed through the collection of longitudinal data tracking candidates as they go from degree through the various career stages.

A newer study on probability of faculty receiving tenure and promotions has found a much more equitable situation. Ginther and Kahn (2006) recently examined three issues with respect to gender differences: (1) the probability of holding a tenure-track job within 5 years of receiving a Ph.D.; (2) for those who hold a tenure track job, the probability of having tenure 11 years after receiving a Ph.D.; and (3) for those who received tenure by 15 years past receipt of a Ph.D., the probability of being a full professor 15 years after receipt of a Ph.D. The study drew on the entire SDR from 1973 through 2001. As summarized in their abstract, the authors found "that in science overall, there is no gender difference in promotion to tenure or full professor after controlling for demographic, family, employer and productivity covariates and that in many cases, there is no gender difference in promotion to tenure or full professor even without controlling for covariates."

The next section presents descriptive data on tenure and promotion, based on data collected in the committee's departmental survey. By examining data on all tenure cases evaluated in the prior 2 years, this analysis avoids the pitfall of studying only men and women who currently hold faculty positions. The following section uses multivariate methods to explore the effect of structural factors on promotion decisions for male and female faculty.

TENURE AND PROMOTION AWARDS

Tenure Descriptive Data

In the case of tenure, the survey first asked whether departments engaged in any tenure decisions during the past 2 academic years (2002-2003 and 2003-2004). Most of the 417 responding departments (78 percent) indicated that

---

[7] See for instance Persell (1983) and McElrath (1992).

they did. Very similar results were obtained by disaggregating the departmental responses by discipline. In all fields, the percentage of departments indicating that they had such tenure cases was between 75 and 84 percent.

Responding departments noted a total of 768 tenure decisions. Most decisions were reported by public institutions (587), rather than private institutions (181). For individual departments that reported any tenure decisions, the median response was two tenure decisions (mean = 2.2), with a range from 1 to 15 decisions. By gender, 125 cases involved female faculty; 642 cases involved male faculty. In 1 case, the gender was not reported. In addition, for 9 cases, the tenure outcome was not reported by departments.

Across all the departments sampled, 15 percent of the tenure candidates were female, compared to 20 percent of the pool of assistant professors, a difference significant at better than .01. There are a number of possible explanations for the smaller percentage of women among tenure candidates compared to the percentage in the tenure pool. If women are more likely than men to resign their position before being proposed for tenure, then we would expect to see fewer women among the tenure candidates. On the other hand, if departments have substantially increased their efforts to hire more women on tenure-track appointments, the disparity may be due to the lag between the time at which a faculty member is hired and the time at which he or she is put up for tenure. Most institutions impose an upper bound on the number of years in which a faculty can serve in a tenure-track position. Early tenure decisions—while not truly rare—are not commonplace either. However, many universities allow for extending the allotted time by up to 2 years to accommodate new parental responsibilities.

The findings on percentages of women among tenure candidates were not uniform across disciplines. Women were most likely to be underrepresented in the fields where they accounted for the largest share of the faculty. Female faculty were considered for tenure in 27 percent of the cases in biology and 15 percent of the cases in chemistry. In both fields, their representation among the assistant professor pool was greater—36 percent of the pool in biology and 22 percent of the pool in chemistry. In the remaining four fields, the differences in representation were less pronounced, although in every case the percentage of women among tenure candidates was less than in the tenure pool. The percentage of women among tenure candidates was 16 percent in civil engineering, 11 percent in electrical engineering, 16 percent in mathematics, and 12 percent in physics. During the same period, the percentage of women among tenure-track assistant professors was 23 percent in civil engineering, 13 percent in electrical engineering, 22 percent in mathematics, and 16 percent in physics.

Contrary to the implication from previous research that the lower percentage of women among tenured relative to untenured faculty results from a lower probability of a positive tenure decision for women, the committee's data showed the opposite. Controlling only for field and gender of the candidate, we found that a woman was marginally more likely than a man to receive tenure ($p = 0.0567$).

TABLE 5-4 Tenure Award Rates by Gender and Discipline

	Male			Female			
Discipline	Granted	Not Granted	Percentage Not Granted	Granted	Not Granted	Percentage Not Granted	Total
All fields	548	85	13.4	115	10	8.0	758
Biology	89	15	14.4	33	5	13.2	142
Chemistry	79	22	21.8	18	0	0.0	119
Civil engineering	76	15	16.5	15	3	20.0	109
Electrical engineering	91	10	9.9	12	0	0.0	113
Mathematics	106	16	13.1	23	1	4.2	146
Physics	107	7	6.1	14	1	6.7	129

NOTE: In 1 case, the gender of the individual up for tenure was unknown, and in 9 cases, the tenure outcome was not reported.
SOURCE: Survey of departments conducted by the Committee on Gender Differences in Careers of Science, Engineering, and Mathematics Faculty.

As shown in Table 5-4, men received tenure in 548 out of 633 cases (87 percent); women received tenure in 115 out of 125 cases (92 percent). (See p. 120 for an explanation of the use of summary survey data.).[8] Disaggregated by field, women had a significantly higher percentage of being granted tenure only in chemistry, where each female faculty member up for tenure was successful. In the other fields, the differences were not significantly different for men and women.

Promotion Descriptive Data

We investigated next whether gender differences exist at the juncture of promotion from associate professor to full professor. Of 411 departments responding to the survey, 70 percent indicated that they had considered a case of promotions to full professor during 2 academic years (2002-2003 and 2003-2004). Over all the fields, 90 percent of men and 88 percent of women proposed for full professor were promoted (see p. 120 for an explanation of the use of summary survey data.). The difference between rates for men and women was not statistically significant, nor were any of the discipline-specific differences shown in Table 5-5.

Most of the 504 cases reported involved public institutions (402), rather than private institutions (106). For individual departments that reported some decisions, the median response was one promotion decision (mean = 2), with a range from 1 to 16 decisions. Disaggregated by gender, 74 cases involved female faculty and 433 cases involved male faculty. In 1 case, the gender was not reported. In addition, among the 508 total cases, the outcome was not reported in 3 cases.

Disaggregated by discipline, female faculty were considered for promotion to full professor in 24 percent of the cases in biology, 14 percent in chemistry, 18 percent in civil engineering, 17 percent in electrical engineering, 9 percent in mathematics, and 7 percent in physics. During the period covered by the faculty survey, the percentage of women among associate professors in the different disciplines was 28 percent in biology, 18 percent in chemistry, 15 percent in civil engineering, 13 percent in electrical engineering, 15 percent in mathematics, and 8 percent in physics. It appears that women are proposed for promotion to the highest academic rank at approximately the same rates at which they are represented among associate professors.

Factors Influencing Tenure and Promotion Decisions

The outcome of a tenure or promotion decision is the product of individual and departmental characteristics. Individual characteristics focus on evaluations of the faculty member's knowledge, skills, and abilities. In the area of research, evaluation may focus on a professor's productivity, measured in terms of publications (i.e., journal articles, books, and chapters), presentations to conferences, or

---

[8] In nine cases involving men who were up for tenure, the outcome was unknown.

TABLE 5-5 Promotion to Full Professor by Gender and Discipline

Discipline	Male			Female			Total
	Promoted	Not Promoted	Percentage Not Promoted	Promoted	Not Promoted	Percentage Not Promoted	
All fields	387	43	10.0	65	9	12.2	504
Biology	50	11	18.0	17	2	10.5	80
Chemistry	65	4	5.8	10	1	9.1	80
Civil engineering	53	6	10.2	10	3	23.1	72
Electrical engineering	69	10	12.7	14	2	12.5	95
Mathematics	68	9	11.7	7	1	12.5	85
Physics	82	3	3.5	7	0	0.0	92

SOURCE: Survey of departments conducted by the Committee on Gender Differences in Careers of Science, Engineering, and Mathematics Faculty.

ability to obtain grants. Teaching evaluations are used as a metric of instructional performance, as is the amount and quality of graduate student supervision. Counts of how many and what kind of university committees and outside professional activities in which a faculty member is involved, and in what capacity, are used to measure service.

Factors affecting these research, teaching, and service performance measures can also have an indirect effect on tenure decisions. Faculty with children, for example, may have less time to pursue research or service activities, and this may reduce a faculty member's chances of being granted tenure. Departments with policies that aid faculty who would otherwise be more negatively affected by family issues—for example, institutions that provide child care or family leave—might mitigate the negative effects of these indirect factors and thereby aid the tenure chances of those faculty members particularly affected by family issues.

Departmental and institutional characteristics also directly affect tenure outcomes. In the most obvious case, both male and female faculty will have lower probabilities of gaining tenure in departments that rarely grant tenure to assistant professors, preferring instead to hire tenured associate or full professors. Different institutions—measured in terms of prestige or type (public versus private)—may grant tenure or promotion at different thresholds. For example, "nationally, about 60 percent of scholars competing for university and college tenure slots gain permanent appointments. At MIT, it is estimated almost 50 percent of the men and women on the tenure track will be invited to make their permanent intellectual home at the Institute."[9]

The committee's survey asked for departments to report institutional characteristics related to individual tenure decisions, but did not ask department respondents to provide information on the individual faculty member beyond their gender and the outcome of the case. Therefore, the model developed here is intentionally underspecified. It does not include likely salient individual factors that influence tenure outcomes. It focuses instead on examining departmental characteristics and policies that might help or hinder female as opposed to male faculty. Factors of particular interest include:

Department size. Larger departments may have more slots available and may therefore provide more opportunities for an assistant professor to advance.

Stopping the tenure clock. Many universities allow faculty to stop or extend the tenure clock if they have a qualifying event, such as the need to care for a family member. Generally, universities limit the number of years that can be added to the period before an assistant professor must be considered for promotion. Either male or female faculty can qualify for delaying the tenure clock. However, use of

---

[9] Anonymous, March 1, 1999, Women and Tenure at the Institute, MIT News Office, available at http://web.mit.edu/newsoffice/1999/trwomen.html. See also Hopkins (2006).

stop-the-tenure-clock policies does extend the period of uncertainty for faculty. In our faculty survey, 78 percent of assistant professors reported that their department or university had a formal family or personal leave policy that allows stopping or extending the tenure clock.

Transparency of tenure and promotion policies. It has been argued that unclear tenure or promotion policies would be particularly detrimental to women if women faculty are less likely to have mentors and obtain information through informal channels. However, evidence collected in the faculty survey indicated that women were at least as well connected to information sources as men. As shown previously, female faculty were more likely than male faculty to have a mentor, and women appear to be as well informed as men about the tenure process (see Appendix 5-1). When asked, 88 percent of both men and women responded that they knew their institution's policy on tenure. However, 81 percent of male faculty but only 75 percent of female faculty responded that they knew their institution's policy on promotion (or knew there was no institutional policy)—which was a significant difference ($p = 0.02$). Most departments use multiple means of informing faculty about tenure policies and procedures: 78 percent of departments reported that the university has written tenure and promotion policies, and 49 percent reported that the department has written procedures.

Departmental culture. Inclusive departments are more likely to pay greater attention to equity. We examined whether departments with more representation of women among the faculty were more or less likely to tenure assistant professors, and whether this varied by gender of the candidate. We also examined whether the percentage of women among untenured assistant professors affected the probability of success of male or female tenure candidates.

Public institutions. Private universities tend to have longer probationary periods than public institutions (NRC, 2001a). Some private institutions prefer to hire junior faculty without tenure and senior faculty with tenure, making it difficult to cross from one status to the other within the institution. Ginther (2001) found that being at a private institution decreased the probability of promotion to tenure. It is less clear whether women's chances of promotion differ relative to those of men at public versus private institutions.

Prestige. Ginther (2001) found evidence that being at a top-ranked university, as defined by rankings provided by the Carnegie Foundation, increased the probability of a promotion to tenure. Note, however, that Ginther's study involved a broader set of institutions than is employed in this study. If top-ranked universities strive to hire only people they expect to tenure, it may be harder to get hired, but easier to gain tenure at such institutions. Conversely, though, some top institutions are less inclined to tenure their own assistant professors.

Multivariate Analysis

Tenure

To explore whether the observed differences between men and women in their success at receiving tenure were statistically significant, and whether some of the variables described above explained those differences, we fitted a generalized linear model to the binary outcome indicating whether a tenure decision was positive or negative. We included various institutional and departmental attributes as explanatory variables in the model, and used the method of generalized estimating equations (GEE) to account for a potential correlation among tenure decisions in the same departments within the same institutions. The explanatory variables included in the model were the following: discipline, gender of the tenure candidate (the variable of interest), prestige of the department, whether the institution was public or private, whether the department allowed faculty to extend the tenure clock for reasons including the arrival of a child, the percentage of tenure-track assistant professors in the department who were female, the percentage of females among the entire faculty of the department, and various two-way interactions between the gender of the tenure candidate and other variables. We did not include in this model any variables that might reflect the productivity of individual faculty members. The reason for this was that the subset of cases with complete information for *all* variables was relatively small.[10]

Results from this analysis are difficult to interpret, at least with regard to gender. While women appeared to be slightly more likely to be promoted and tenured than men, the effect of gender on tenure decision must be interpreted cautiously. This is because the interaction between the gender of the candidate and the percentage of females in the tenure-track pool was also evident. Women appeared to be more likely to be promoted when there was a smaller percentage of females among tenure-track faculty. Therefore, the difference between women and men in their tenure success was more pronounced in departments with fewer women assistant professors. After accounting for all the avenues through which gender affects tenure, across all fields, 93 percent of women and 83 percent of men who were considered for tenure were successful.

Assistant professors (both male and female) were significantly more likely to receive tenure at public institutions, where 92 percent of those considered became tenured, than at private institutions, where 85 percent gained tenure ($p = 0.029$). The probability of gaining tenure was greater in departments of lower ($p = 0.017$) or medium ($p = 0.073$) prestige compared to those in the highest prestige category.

Because the presence of the interaction between the gender of the candidate and the percentage of women among tenure-eligible faculty prevented us from

---

[10] See Appendix 5-2 and 5-3 for detailed tables.

interpreting the impact of either gender or percentage of females among assistant professors on tenure decisions, we did not attempt to untangle other associations. Figures 5-1 and 5-2 show the estimated probability of a positive tenure decision for men and women as a function of the percentage of tenure-eligible faculty who are female and as the proportion of female faculty in the department. To compute the probabilities in Figure 5-1, we held all other factors constant. Similarly, to compute the probabilities in Figure 5-2, we held the percentage of women among tenure-eligible faculty constant at 10 percent (two outer curves) or at 50 percent (two inner curves) for men and women.

Discipline, stop-the-clock policies, and overall departmental size were not associated with the probability of a positive tenure decision for either male or female faculty.

As a final comment, we note that when an interaction between a discrete covariate and other covariates in the model is present and the outcome variable is discrete (as is the case in our logistic regression model for tenure decision), unequal residual variances in each of the levels of the discrete covariate can have a profound effect on inference. Unequal group variances inflate the size of the estimated regression coefficients, thus introducing a bias in predictions relevant to differential outcomes for men and women. Therefore, trying to determine the effect of a covariate on, for example, male and female faculty cannot be done in

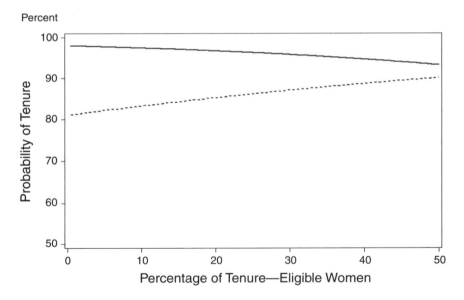

FIGURE 5-1 Probability of tenure for male and female candidates as a function of the percentage of tenure-eligible women in the department. Solid line corresponds to women and dotted line to men.

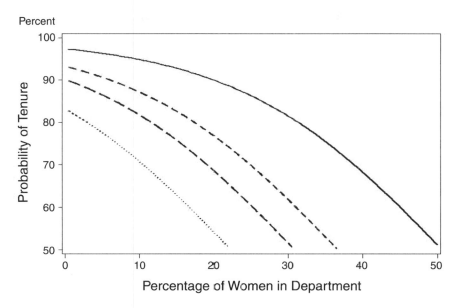

FIGURE 5-2 Probability of tenure for male and female candidates as a function of the percentage of women in the department. The solid line corresponds to female candidates when the percentage of tenure-eligible faculty who are women is 10 percent. The dotted line corresponds to men. The two inner curves correspond to women (upper) and men (lower) when the percentage of tenure-eligible faculty who are women is 50 percent.

the usual manner. The class of models known as heterogeneous discrete choice models (e.g., Alvarez and Brehm, 1995[11]) has been proposed for analysis of this type of data.

Promotion

We again fitted a generalized linear model to the binary outcome indicating the promotion decision and accounted for possible correlation between cases in the same department within the same institution by implementing the method of GEE to obtain improved standard errors for all model parameters. Most of the explanatory variables in the promotion decision model were the same as those used earlier in the tenure decision model. Here, however, we considered the percentage of women among tenured associate professors to be the "promotion pool."

---

[11] Alvarez, R.M. and J. Brehm, 1995, American ambivalence towards abortion policy: development of a heteroskedatic probit model of competing values, *American Journal of Political Science,* 39, 1055-1089.

None of the variables included in the model appeared to be associated with the probability that a candidate was promoted to full professor. In particular, there were no significant differences ($p = 0.607$) in the probability of promotion to full professor due to gender of the candidate, after accounting for the other potentially important factors. Therefore, it appears that once proposed for promotion to full professor, women and men fare about the same across all types of institutions and departments.

There are several reasons our analyses may produce different results than those reported previously in the literature. First, the studies use different samples; for example, faculty at different types of institutions or in different fields. A more interesting difference is conceptual. While the committee's survey focused on the tenure and promotion *decision*, most prior studies focus on whether or not an individual is tenured or promoted by a particular point in time. To better illustrate this distinction, one can compare the results of a 2006 Pennsylvania State University study of faculty cohorts entering the tenure track between 1990 and 1998 As Table 5-6 shows, 7 years after entering the tenure track, the percentage of men who were tenured professors at Penn State was higher than the percentage of women who were tenured.

However, in a second analysis, Dooris, Guidos, and Miley (2006) examined the outcome of reviews of faculty who were evaluated 6 years after being hired. As seen in Table 5-7, the rates for men and women were not significantly different at the sixth year review, with observed values of 90 percent for men and 87 percent for women ($p = 0.69$). The differences between these two ways of examining the data may be due to the departure from Penn State by some faculty, who never came up for review. Alternately, the results may reflect that some faculty took leave, delaying the tenure decision for them beyond the sixth year.

These two foci—tenure status after a specific time period and tenure decisions—correspond to different but partially overlapping groups of faculty. In the committee's study, the denominator included any faculty who came up for a tenure or promotion decision. In the other studies briefly surveyed above, the denominator included both individual faculty who came up for a decision and tenure-track faculty who have not yet reached that point. A second reason is that men and women may spend different amounts of time at each rank. This topic is discussed below.

TIME IN RANK

Although women are as successful as men when they are considered for tenure, differences in gender distributions at different faculty ranks may relate to differences in how long men and women spend within ranks. In general, the literature suggests women take longer to get tenure or a promotion. According to one study, across all fields (S&E and non-S&E) except for engineering and mathematics/statistics, women wait longer to attain tenure. Significant differences

TABLE 5-6 Tracking Cohorts Entering the Tenure Track Through 7 Years: Pennsylvania State University

Cohort Year	All Entrants			Female			Male		
	Entrants	Tenured	Rate (%)	Entrants	Tenured	Rate (%)	Entrants	Tenured	Rate (%)
1990	121	70	58	40	19	48	81	51	63
1991	93	55	59	30	15	50	63	40	63
1992	151	89	59	55	28	51	96	61	64
1993	103	55	53	31	12	39	72	43	60
1994	134	63	47	50	17	34	84	46	55
1995	127	70	55	53	30	57	74	40	54
1996	91	45	49	29	12	41	62	33	53
1997	160	87	54	52	25	48	108	62	57
1998	183	107	58	75	38	51	108	69	64
Totals	1163	641	55	415	196	47	748	445	59

SOURCE: Dooris, Guidos, and Miley, 2006: Table 1.

TABLE 5-7 Results of Faculty Reviews at 2, 4, and 6 Years Following Hire: Pennsylvania State University, 2004-2005

Year	Total			Men			Women		
	2nd	4th	6th	2nd	4th	6th	2nd	4th	6th
No. of cases reviewed	113	140	107	63	86	60	50	54	47
No. with continuation recommended	109	121	95	61	75	54	48	46	41
No. forwarded for early tenure	5	13	n/a	3	9	n/a	1	4	n/a
Percent with positive recommendation	96	86	89	97	87	90	96	85	87

SOURCE: Dooris, Guidos, and Miley, 2006: Table 3.

in which men were favored were found in the biological sciences and psychology and the social sciences. In engineering, however, women were significantly more likely to receive tenure first (Astin and Cress, 2003). A separate study of physician faculty of U.S. medical schools found that women were "much less likely than men to have been promoted to associate professor or full professor rank after a median of 11 years of faculty service" (Tesch et al., 1995). Finally, Kahn (1993) found that for academic economists, the time between receipt of Ph.D. and tenure for men was 7 years, while for women it was 10 years. Data for individual universities also show this trend at such schools as the University of California, Berkeley, MIT, and Duke University (NAS, NAE, and IOM, 2007).

Data from the 2004 National Study of Postsecondary Faculty show that, among faculty who earned their doctorates in U.S. institutions and were employed full time in 1997 in academic institutions in biology, physical science, engineering, and mathematics, women averaged 9 years as assistant professors, compared to an average of 7.6 years spent by men, though this difference was not statistically significant at the 0.05 level (see Table 5-8, and Appendixes 5-4, 5-5). Female full professors were promoted to that rank an average of 13 years after first being hired, compared to an average of 10.1 years for men, which is statistically significant at the 0.05 level. Thus, the gap between men and women in years between first hire and most recent promotion grows between the associate and full professor ranks.

Consistent with the data on average time in rank, the NSOPF data showed a greater percentage of female associate professors (16.8 percent) spent 11 to 15 years as assistant professors, compared to 8.4 percent of male associate professors (Appendix 5-3), although this was not statistically significant at the 0.05 level. Twenty percent of male associate professors were promoted to that rank after 5 or fewer years as an assistant professor, compared to 13.1 percent of female associate

TABLE 5-8 Mean Number of Years Between Rank Achieved and First Faculty or Instructional Staff Job, by Gender, for Full-Time Faculty at Research I Institutions, Fall 2003

	Years Between Current Rank Achieved and Employment Start at Postsecondary Institutions	
	Associate Professor Mean (Std. error)	Full Professor Mean (Std. error)
Total Faculty	7.8 (0.3)	10.4 (0.2)
Men	7.6 (0.3)	10.1 (0.2)
Women	9.0 (0.9)	13.0 (0.6)

NOTE: Numbers are for full-time faculty with instructional duties for credit, teaching biology, physical sciences, engineering, mathematics, or computer science.
SOURCE: NCES, NSOPF: 2004 National Study of Postsecondary Faculty March 30, 2006.

professors; although again, this was not statistically significant at the 0.05 level. The distribution of years spent in the associate professor rank shows 36.3 percent of men were promoted to full professor after 10 years or less, compared to 26.4 percent of women (see Appendix 5-5).[12]

The committee's faculty survey differs in some respects from the NSOPF:04 data in that it includes only faculty at RI institutions and does not exclude faculty who earned their doctorates outside the United States. The sample size of 634 used to construct Table 5-9 reflects a loss of about 50 percent from the original sample of about 1,250 respondents, because information on time in rank could only be calculated for those faculty members who received at least one promotion at their current institution. Because neither sampling weights nor nonresponse weights were used, care should be taken in generalizing the results to the population of all faculty.

Despite the differences in samples, the committee's survey found results similar to the NSOPF:04 study for time in rank. Table 5-9 presents data on the mean number of months that faculty who were promoted to associate professor in each of the six disciplines surveyed spent in the rank of assistant professor. Similar calculations were made for male and female full professors. Across the six comparisons for faculty who were currently associate professors, women averaged a significantly longer time in rank in all fields except civil engineering and electrical engineering, where women's time in rank was not different from men's. For current full professors, women spent significantly longer time in the rank of assistant professor in all disciplines, and in three disciplines, it was statistically significant.

It is interesting to note that the average number of months spent as an assistant professor has been rising over time, as indicated by the longer durations for both male and female associate professors, as compared to their counterparts who were promoted at an earlier time period and are now full professors.

The measure used in Table 5-9 does not include years spent as a postdoc, employed outside of academia, or unemployed. We also calculated the time that elapsed between the date of obtaining a Ph.D. and the date of promotion to associate professor with tenure, shown in Table 5-10. This second measure accounts for the time spent in one or more postdoctoral positions prior to the first tenure-track job. It shows the greater number of months to promotion to associate professor with tenure (an average of 95.0 [see Table 5-10] months compared to 68.6 months [see Table 5-9] spent as an assistant professor), with trends over time and contrasts by gender varying from those reported in Table 5-9. The number of months between receipt of Ph.D. and promotion to associate professor with tenure shows greater increases over time than the measure of time spent as an assistant professor, reflecting the increased prevalence and duration of postdoctoral appointments.

---

[12] In the NSOPF data, there are many more men in the sample than women and the standard errors for women are much larger.

TABLE 5-9 Mean Number of Months Spent as an Assistant Professor

Discipline	Current Associate Professors		Current Full Professors	
	Men	Women	Men	Women
Biology	68 (4) (21)	74 (3) (36)	63 (3) (25)	68 (4) (28)
Chemistry	62 (4) (27)	72 (2) (33)	56 (3) (43)	67 (3) (22)
Civil engineering	69 (4) (29)	69 (2) (30)	61 (4) (16)	65 (4) (19)
Electrical engineering	64 (6) (13)	67 (3) (23)	55 (4) (17)	58 (3) (21)
Mathematics	40 (6) (18)	60 (3) (30)	46 (4) (36)	47 (4) (22)
Physics	55 (3) (27)	60 (3) (33)	55 (2) (41)	61 (4) (24)

NOTES: The first set of parentheses indicates standard error of the mean, and the second set of parentheses denotes number of observations used in the calculation.

There were only 634 faculty with current rank as associate or full professor who were hired at their current institution as tenure-track assistant professors, who work full time, and who have a Ph.D. Only those faculty who were promoted to associate with tenure from assistant were used in the calculations. We omitted departments that did not provide information on gender of faculty. The low average computed for men in mathematics who are currently associate professors is influenced by a reported time of an assistant professor of only 4 months. If we eliminate that record from the data, the new average as associate is 50 months, more similar to the average time computed for women associate professors. We also omitted one outlier who reported being unemployed for 27 years following graduation, three individuals with negative time to promotion (promotion happened before hire), and three individuals who spent over 320 months (over 26 years) as assistant professors.

SOURCE: Faculty survey conducted by the Committee on Gender Differences in Careers of Science, Engineering, and Mathematics Faculty.

Although men in the full professor cohort generally experienced fewer months between receiving their Ph.D. and being promoted to associate professor with tenure than women, the results were very mixed for the sample of current associate professors.

Turning next to the promotion to full professor, Table 5-11 presents weighted means of time spent as an associate professor for the 311 full professors for whom data were available. In contrast to the NSOPF data, women who were currently full professors spent significantly more time as associate professors in chemistry, mathematics, and electrical engineering, where the differences between men and women were not significantly different at the 5 percent level. Overall, the data were not clear for both full and associate professors.

Multivariate Modeling of Time in Assistant Professor Rank

A Cox proportional hazards model[13] was fit to the measure of time in rank as assistant professor. A nearly identical model was fit to the data on time elapsed

---

[13] Cox, D.R. and Oaks, D., 1984, *Analysis of Survival Data*, London: Chapman & Hall.

TABLE 5-10 Mean Number of Months Between Receipt of Ph.D. and
Promotion to Associate Professor

Discipline	Current Associate Professors		Current Full Professors	
	Men	Women	Men	Women
Biology	158 (8) (13)	135 (5) (27)	102 (6) (13)	122 (48) (20)
Chemistry	112 (4) (20)	127 (9) (23)	95 (3) (31)	88 (5) (17)
Civil engineering	113 (6) (17)	80 (4) (9)	54 (4) (4)	68 (5) (4)
Electrical engineering	110 (14) (4)	77 (2) (11)	78 (10) (11)	85 (4) (10)
Mathematics	66 (9) (16)	101 (10) (25)	80 (3) (32)	88 (4) (19)
Physics	134 (5) (20)	113 (6) (20)	105 (3) (34)	104 (6) (14)

NOTES: The first set of parentheses indicates standard error of the mean, and the second set of parentheses denotes number of observations used in the calculation.

There were only 418 faculty with current rank as associate or full who were hired at their current institution as tenure-track assistant professors, who work full time, and who have a Ph.D. Only those faculty who were promoted to associate with tenure from assistant were used in the calculations. We omitted departments who did not provide information on gender of faculty, as well as one outlier who reported being unemployed for 27 years following graduation, three individuals with negative time to promotion (promotion happened before hire), and one person who spent 321 months as assistant professor. The numbers of cases used in Table 5-9 and here differ because we did not have reliable information about the time of graduation for some faculty. There were also several outliers. For example, a female math associate professor reported that 307 months (over 25 years) elapsed between obtaining her Ph.D. and her promotion to associate professor with tenure. A male math associate professor reported only 5 months elapsed between obtaining his Ph.D. and his promotion to associate professor with tenure.

SOURCE: Faculty survey conducted by the Committee on Gender Differences in Careers of Science, Engineering, and Mathematics Faculty.

between receipt of Ph.D. and promotion to associate professor, except the second model could not include a measure of academic age, defined as time elapsed between Ph.D. and hire as tenure-track assistant professor. Other variables in both models included gender, discipline, current rank, an indicator for whether family leave was taken, type of institution (public or private), prestige, percentage of women among faculty in the department, and various two-way interactions with gender.

Results suggest that there is a complex interplay among the various factors in the model and time in rank as assistant professor. Only two of the factors—type of institution and the percentage of women among departmental faculty—appeared to have no significant association with time in rank as assistant professor. All other factors, including the interactions between gender and current rank, gender and academic age, and gender and prestige of the institution were significantly associated with time in rank as assistant professor.

Because of the presence of significant interactions, it is difficult to provide an interpretation of the effect of the main factors. In Table 5-12, we present some of

TABLE 5-11 Mean Number of Months Spent as an Associate Professor

Discipline	Men			Women		
	Mean	Std. error	n	Mean	Std. error	n
Biology	75	4	25	74	6	28
Chemistry	62	4	42	80	7	21
Civil engineering	72	7	16	75	9	18
Electrical engineering	61	8	17	70	4	20
Mathematics	59	4	35	68	5	22
Physics	77	5	41	79	7	25

NOTE: Table entries are computed using faculty who are currently full professors and who were promoted from associate to full and from assistant to associate at their current institution. We used only full-time faculty with a Ph.D. for whom we had complete information about the time of both promotions, resulting in a total of 311 observations. The proportion of nonrespondents was similar among men and women for all ranks.
SOURCE: Faculty survey conducted by the Committee on Gender Differences in Careers of Science, Engineering, and Mathematics Faculty.

the results obtained when fitting the Cox proportional hazards regression model and summarize findings via figures that show the probability of promotion to associate professor at each time point for men and women in different disciplines, who are of different current ranks and at universities of different prestige. Overall, it appears that women take significantly longer to achieve promotion to associate professor with tenure, but this gender effect is confounded with current rank, discipline, and various other factors.

TABLE 5-12 Results Obtained from a Cox Proportional Hazards Regression Analysis of 351 Cases That Had Complete Promotion and Covariate Information

Effect	p-value	Hazard Ratio
Gender (1 = M, 2 = F)	0.007	0.360
Current rank (1 = F, 2 = A)	<0.0001	0.166
Academic age	<0.0001	1.046
Academic age × gender	<0.0001	0.981
Rank × gender	0.001	2.282
Prestige	0.007	0.483
Prestige × gender	0.001	1.744
Family leave (0 = no)	<0.0001	0.192
Biology vs. civil engineering	0.307	0.709
Chemistry vs. civil engineering	0.046	1.778
Mathematics vs. civil engineering	<0.0001	4.126
Electrical engineering vs. civil engineering	<0.0001	4.927
Physics vs. civil engineering	0.154	1.511

A hazard ratio below 1 indicates individuals in the category with the higher value of the explanatory variable "survive" longer. In this case, a faculty member "survives" in the rank of assistant professor if, in the next month, he or she does not get promoted to associate professor with tenure. For example, in the absence of interactions between gender and other variables, we would have concluded the average "risk" a female faculty will be promoted to associate professor with tenure at a given time point (given that she had not been promoted up until that time) is about 36 percent of that of a male. However, the presence of significant interactions prevents us from drawing conclusions about the effects of gender, rank, prestige, academic age, and others individually.

Figures 5-3 (a-d) show the survival curves for men and women who are currently associate or full professors in biology at high-prestige institutions (Figure 5-3a) or at medium-prestige institutions (Figure 5-3b). Figures 5-3c and 5-3d show the corresponding survival curves for faculty in electrical engineering. Male full professors are represented by a gray solid curve and male associate professors are represented by a gray dotted curve. Female full professors are represented by a black solid curve and female associate professors are represented by a black dotted curve.

The plots shown in Figures 5-3 (a-d) reflect some of the complexities in the relationship between time in rank as assistant professor, current rank, gender, and prestige of the institution. For example, consider first biology. We note that at high-prestige institutions and at any time point t + 1, a male who was currently a full professor had a higher chance of getting promoted and tenured than a male

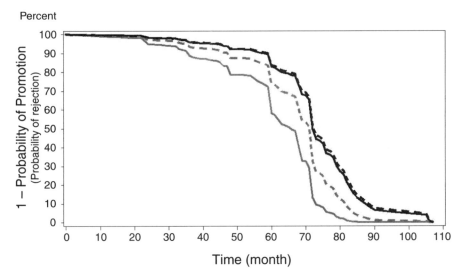

FIGURE 5-3(a) Survival curves in biology at highest prestige institutions.

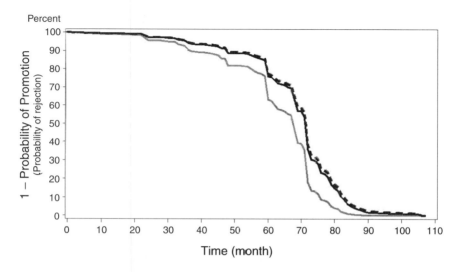

FIGURE 5-3(b) Survival curves in biology at medium-prestige institutions.

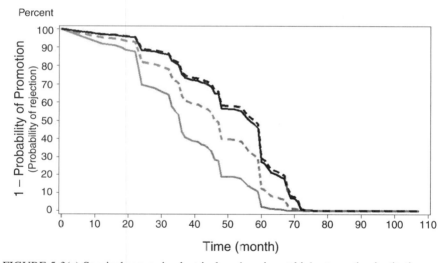

FIGURE 5-3(c) Survival curves in electrical engineering at highest prestige institutions.

who was currently an associate professor, and in turn, they both had a higher chance of promotion at month t + 1 (given that they had not been promoted until then) than a woman who was currently a full professor or a woman who was currently an associate professor. These differences vanish, however, if we consider institutions of medium prestige. In that case, while a man who was currently a full

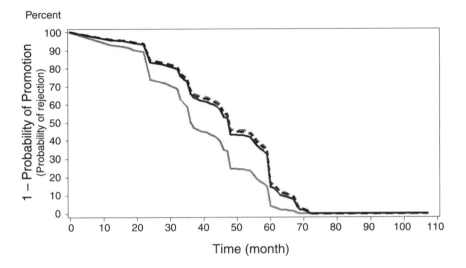

FIGURE 5-3(d) Survival curves in electrical engineering at medium-prestige institutions.

professor still had a higher chance of getting promoted at any time, given he has not been promoted earlier, there were no differences between faculty who were currently associate professors or women who were currently full professors. In the case of electrical engineering, we observed a similar pattern even though the probability of promotion increased to one at a faster rate.

Academic age (time between receipt of Ph.D. and hire as tenure-track assistant professor) was negatively and significantly associated with time as assistant professor: The longer the time elapsed between Ph.D. and hire, the shorter the time spent in rank as assistant. This finding is consistent with the greater publication record faculty who have spent time as postdocs can present at the time of a tenure review. Academic age may contribute to the gender differential seen in the simple means of time in rank by gender, since the effect of academic age is significantly stronger for men than for women ($p < 0.0001$). This greater impact may reflect that men may be more likely than women to spend time after receiving their doctorate and prior to taking their first academic job pursuing professional activities, such as postdoctoral research.

Another important factor affecting time in rank as assistant professor is the increasingly available option to take family leave and stop the tenure clock. Our results show a very significant effect of stopping the tenure clock (p-value < 0.0001; see Table 5-12). The "risk" of promotion of a faculty member who stopped the tenure clock is only about 80 percent of the "risk" of promotion of a faculty member who did not, given that neither had been already promoted at a given time. Consider, as an example, two faculty members with similarly impressive academic credentials so that their "risk" of promotion becomes one if enough time has elapsed since hiring. If one of them takes a 1-year leave at the beginning of his or her probationary period, then he or she will lag behind the person who did not take the leave with respect to promotion status, but the difference in the "risk" of promotion will get smaller and smaller as the overall probabilities of promotion for *both of them* become larger. The effect of this factor was similar for both men and women. However, our data confirm that women were more likely to take family leave. Table 5-13 shows that 10.2 percent of female and 6.4 percent of male associate professors stopped the tenure clock. Also, stopping the clock is becoming more common over time. Virtually no faculty who are currently full professors stopped the clock, but among assistant professors, 19.7 percent of women and 7.4 percent of men have already stopped the clock. These percentages are likely to continue growing in the future.

One question our survey does not permit addressing is whether a faculty member who stopped the tenure clock has a decreased probability of promotion. To answer that question we would need a longitudinal study where faculty can be followed from the time they were hired until the time they were promoted. Our survey, which collected a snapshot cross-sectional set of data, is not appropriate for this type of question.

TABLE 5-13 Number of Faculty by Gender and Rank Who Reported Stopping or Not Stopping the Tenure Clock or Who Did Not Respond to the Survey Question

Gender and Rank	Stopped Clock	Did Not Stop Clock	Nonrespondent	Total
Male full professor	1	52	261 (83)	314
Female full Professor	2	46	18 (80)	237
Male associate professor	14	137	68 (31)	219
Female associate professor	29	184	71 (25)	284
Male assistant professor	17	211	2 (0.8)	230
Female assistant professor	56	226	2 (0.7)	284

NOTES: Numbers in parentheses are percentage of nonrespondents in each group.
Only full-time faculty with a Ph.D. and with the rank of assistant, associate, or full professor were used in the calculation. There were 1,568 such faculty. Note that many of these individuals are missing information on other variables, and thus this table includes many more persons than most of the other tables in Chapter 5.
SOURCE: Faculty survey conducted by the Committee on Gender Differences in Careers of Science, Engineering, and Mathematics Faculty.

Men who are full professors today spent the least time in rank as assistant professors. This is true across all disciplines, prestige of institution, and other factors. Whether males who are currently associate professors have spent more or less time in rank as assistant than women who are currently full professors depends on the institution and discipline. It is probably fair to state that women who are currently associate professors have spent the longest time in rank as assistant professors in most cases.

Faculty in biology, physics, and civil engineering are similar in terms of time in rank as assistant professor. In chemistry, math, and electrical engineering, the time to promotion to associate professor was similar and significantly shorter. The difference between disciplines was similar for both genders. There were no significant differences between private and public institutions once all other effects were accounted for.

Results for the measure of time elapsed between award of Ph.D. and promotion to associate professor with tenure were different and easier to interpret from the results discussed above. Using this measure, the time in rank as assistant did not differ between men and women (although it took women slightly longer to be promoted to associate from the time of graduation with a Ph.D.), and it did not differ across institutions of different prestige. Time elapsed between Ph.D. and promotion to associate was highest for faculty who were currently associate professors (as before) and for faculty in biology relative to the other disciplines.

Multivariate Modeling of Time in Associate Professor Rank

To examine what institutional and individual characteristics influence the number of months full professors in our sample spent as associate professors before being promoted, we examined data on 265 respondents. It was necessary to limit the sample to those full professors who had remained at the same institution since they were hired as assistant professors in order to obtain relevant data on institutional characteristics and policies. The sample does not include 20 cases who reported first being promoted to associate professor without tenure and then to associate professor with tenure. The attrition in the analysis sample due to data constraints limits the generalizability of the results to faculty who progressed from assistant professor to associate professor with tenure and then to full professor at the same institution.

Time in rank as associate—computed as the difference in months between first promotion to associate with tenure and promotion to full professor—was modeled as a function of individual characteristics (including gender, discipline, and academic age) and institutional characteristics (including public/private university, prestige, tenure clock policy, and percent of female faculty in the department). All two-way interactions with gender were also estimated. We again used a Cox proportional hazards regression model to explore the association between time in rank as associate and institutional and individual attributes.

Overall, there was no significant difference between male and female faculty in the time spent as an associate professor. Faculty (both male and female) at the higher prestige institutions spent longest in rank as associate professors, while males at the lowest-prestige institutions received promotion earliest. For example, in biology, the probability of promotion after about 8.5 years in rank as associate professor was approximately 80 percent at institutions of highest prestige for both men and women. At institutions of lower prestige, about 80 percent of the men were promoted after 5 years in rank as associate, while 6.8 years elapsed before 80 percent of the women at the lowest prestige institutions received promotion to full professor. Women in universities ranked in the bottom two tertiles spent about the same amount of time in the associate rank. There were no statistically significant differences across disciplines or between public and private institutions. Academic age was positively associated with time in rank.

Figure 5-4 shows the (conditional) probability of promotion to full professor at month t + 1 given that no promotion had occurred until month t. The six curves correspond to prestige of the institution (highest = light gray, middle = dark gray, lowest = black) and to gender (solid = male, dotted = female). Figure 5-4a was drawn for biology at a private institution with 17 percent female faculty, and Figure 5-4b was drawn for electrical engineering.

Figures 5-4 (a-b) show the (conditional) probability of promotion to full professor at month t + 1 given that no promotion had occurred until month t. The six curves correspond to prestige of the institution (highest = light gray, middle = dark gray, lowest = black) and to gender (solid = male, dotted = female). Figure

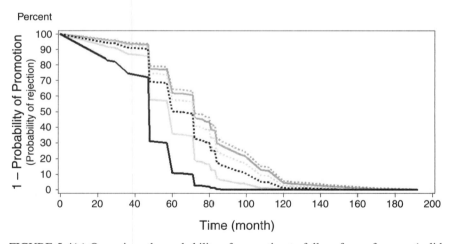

FIGURE 5-4(a) One minus the probability of promotion to full professor for men (solid curves) and women (dashed curves) in biology. Light gray denotes institutions of highest prestige, dark gray represents institutions of medium prestige and black represents institutions of lower prestige.

Percent

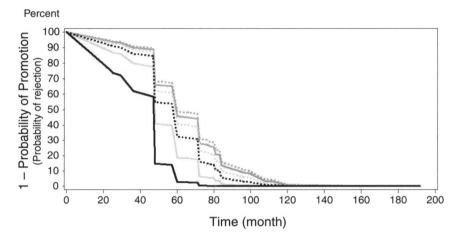

FIGURE 5-4(b) One minus the probability of promotion to full professor for males (solid curves) and females (dashed curves) in electrical engineering. Light gray denotes institutions of highest prestige, dark gray represents institutions of medium prestige and black represents institutions of lower prestige.

5-4a was drawn for biology at a private institution with 17 percent female faculty, and Figure 5-4b was drawn for electrical engineering.

SUMMARY OF FINDINGS

The survey results yielded some surprising findings about the award of tenure, promotion to full professor, and time and rank for female and male faculty members.

Award of Tenure

Finding 5-1: In every field, women were underrepresented among candidates for tenure relative to the number of female assistant professors. Most strikingly, women were most likely to be underrepresented in the fields in which they accounted for the largest share of the faculty—biology and chemistry. In biology and chemistry, the differences were statistically significant. In biology, 27 percent of the faculty considered for tenure were women, although women represented 36 percent of the assistant professor pool. In chemistry those numbers were 15 percent and 22 percent, respectively. This difference may suggest that female assistant professors were more likely to leave before being considered for tenure than were men. It might also reflect increased hiring of female assistant professors in recent years (compared with hiring 6 to 8 years ago). Note, however, that the probability of representation in the tenure pool in a cross-sectional study such as this is completely confounded with time.

Finding 5-2: Given that the interaction between the gender of the candidate and the percentage of women in the tenure-track pool was statistically significant ($p = 0.012$), women appeared to be more likely to be promoted when there was a smaller percentage of women among the tenure-track faculty, resulting in a greater difference between men and women in their tenure success in departments with fewer female assistant professors. (Figures 5-1 and 5-2 and Appendix 5-3)

Finding 5-3: Women were more likely than men to receive tenure when they came up for tenure review. When controlling only for field and gender of the candidate, we found that women were marginally more likely than men to receive tenure ($p = .0567$). Women received tenure in 92 percent of the cases (115 out of 125) compared to 87 percent of the cases for men (548 out of 633). (Table 5-4)

Finding 5-4: Discipline, stop-the-tenure-clock policies, and departmental size were not associated with the probability of a positive tenure decision for either male or female faculty members who were considered for tenure. Both male and female assistant professors were significantly more likely to receive tenure at public institutions (92 percent) compared to private institutions (85 percent; $p = 0.029$). (Appendix 5-2)

Finding 5-5: Eighty-eight percent of both male and female survey respondents stated that they knew their institution's policy on tenure. Eighty-one percent of male faculty knew their institution's policies on promotion. However, only 75 percent of female faculty respondents knew their institution's policy on promotion, which is statistically significant ($p = 0.02$). (Appendix 5-1)

Promotion to Full Professor

Finding 5-6: For the six disciplines surveyed, 90 percent of the men and 88 percent of the women proposed for full professor were promoted—a difference that was not statistically significant. There was no significant difference in the probability of promotion to full professor due to gender of the candidate, after accounting for other potentially important factors such as disciplinary differences, departmental size, and use of stop-the-tenure-clock policies. Once proposed for promotion to full professor, women and men appear to have fared about the same across all types of institutions and departments. (Table 5-5)

Finding 5-7: Women were proposed for promotion to full professor at approximately the same rates as they were represented among associate pro-

fessors. Female faculty in biology were considered for promotion in 24 percent of the cases (28 percent of the associate professor pool); 14 percent of the cases in chemistry (18 percent of the pool); 18 percent of the cases in civil engineering (14 percent of the pool); 17 percent of cases in electrical engineering (13 percent of the pool); 9 percent of cases in mathematics (15 percent of the pool); and 7 percent of the cases in physics (8 percent of the pool). (Table 5-5)

Time in Rank

Finding 5-8: Time in rank as an assistant professor has grown over time for both male and female faculty. Men who were full professors at the time of the survey had spent the least amount of time in rank as assistant professors. This was true across all disciplines.

Finding 5-9: Women who were associate professors at the time of the survey had averaged a significantly longer time in rank as assistant professors in all fields except electrical engineering, where women's shorter time in rank was not significantly different ($p = 0.999$). It is difficult to determine whether these apparent differences persist once we control for individual and departmental characteristics such as length of postdoctoral experience and stopping the tenure clock for family leave. While women did appear to remain at the rank of assistant professor longer than men, the differences between genders depended upon factors such as the prestige of the institution, the time elapsed since the completion of the doctoral degree, and the current rank of the individual. Both male and female faculty spent longer time in the assistant professor ranks at institutions of higher prestige. (Table 5-9)

Finding 5-10: Male and female faculty who stopped the tenure clock spent significantly more time as assistant professors than those who did not (an average of 74 months compared to 57 months). They had a lower chance of promotion to associate professor (about 80 percent) at any time (given that they had not been promoted until then) than those who did not stop the clock. Everything else being equal, however, stopping the tenure clock did not affect the probability of promotion and tenure; it just delayed it by about a year and a half. It is unclear how that delay affected female faculty, who were more likely than men to avail themselves of this policy. Although the effect of stopping the tenure clock on the probability of promotion and tenure was similar for both male and female faculty, 19.7 percent of female assistant professors in the survey sample availed themselves of this policy compared to 7.4 percent of male assistant professors. At the associate professor level, 10.2 percent of female faculty compared to 6.4 percent of male faculty stopped the tenure clock. (Table 5-13)

Finding 5-11: There is no significant difference between male and female faculty in terms of the time spent as an associate professor. Time in rank as associate professor is significantly associated with the prestige of the institution. Faculty at lower prestige institutions tend to be promoted to full professor earlier than those at the highest prestige institutions. (See Figures 5-4a and 5-4b for examples in biology and electrical engineering.)

Time from Receipt of Ph.D.

Finding 5-12: Overall, it appears that women faculty took significantly longer from receipt of Ph.D. to promotion to associate professor with tenure, but this gender effect was confounded with current rank, discipline, and other factors. It is difficult to determine whether these apparent differences persist once we control for individual and departmental characteristics such as length of postdoctoral experience and stopping the tenure clock for family leave. While women did appear to remain at the rank of assistant professor longer than did men, the differences between gender depended on factors including the prestige of the institution, the time elapsed since completion of the doctoral degree, and the current rank of the individual. (Table 5-12)

Finding 5-13: The longer the time elapsed between receipt of the Ph.D. and hire as an assistant professor, the shorter the time spent in rank before gaining tenure. Academic age may contribute to the gender differential seen in the simple means of time in rank by gender, since the effect of academic age was significantly stronger for men than for women ($p < 0.0001$).

Our findings, which focus on the tenure and promotion decisions themselves rather than the proportions of tenured women or female full professors, differ from previous studies that indicated women fare worse than men, both in receiving tenure and in being granted a promotion. It does appear that women spend longer in assistant professor positions than men, but the complex interplay between different factors and the small number of cases for analysis limit the extent to which we can state that gender is associated (or not) with time in rank.

This study's findings on the success of female faculty in obtaining tenure may relate to the particular focus on scientists and engineers at the most research-oriented universities. Alternatively, these results may reflect an improved climate for women scientists and engineers in RI institutions, given that our data examine a relatively recent period (2002-2003 and 2003-2004).

The findings on women's relative success in the tenure decision process relate importantly to our findings on time in rank. The greater time in rank as assistant professor among female faculty who are currently associate professors compared to men can be partly attributed to women's greater use of stop-the-tenure-clock policies.

The committee interprets these data as indicating that stop-the-tenure-clock policies allowed women who are currently associate professors sufficient flexibility to both assemble a strong tenure case and tend to family responsibilities. In contrast, the cohort who are currently full professors did not benefit from the recently enacted stop-the-tenure-clock policies, and thus these women would have had to meet the same timetable as men, despite their greater family caretaking responsibilities. If they had not met the same deadlines they would not have been granted tenure and promotion and would not now be found among the ranks of full professors.

The growth in time in rank for assistant professors may be attributable in part to increasing expectations about scholarly productivity, reinforcing the need for women to avail themselves of family leave if they are to successfully earn tenure. Stop-the-tenure-clock policies, which are taken advantage of by both male and female faculty, further increase the average length of time faculty spend as assistant professors and thereby extend the period of uncertainty for these faculty.

These findings suggest that there have been major changes over time in women's opportunities to succeed in academic careers. If some of these changes can be attributed to changes in university policies such as the stop-the-tenure-clock policy for family care, this is good news. It suggests that universities can change long-established policies that might have prevented one group of scientists and engineers from advancing to permanent careers within the institution. It also opens the door to considering other established university policies that may hinder our country's ability to profit from creativity of all trained scientists, both male and female. For example, one policy that might be opened for reexamination is the usual requirement that all assistant professor appointments be full time. Part-time appointments would allow both women and men the opportunity to better balance family and career over time.

6

Key Findings and Recommendations

The surveys of academic departments and faculty have yielded interesting and sometimes surprising findings. **For the most part, male and female faculty in science, engineering, and mathematics have enjoyed comparable opportunities within the university, and gender does not appear to have been a factor in a number of important career transitions and outcomes.** Where these findings document real changes in university policies, such as the stop-the-tenure-clock policy for family care, this is good news. It suggests that universities can change long-established policies that might have prevented one group of scientists and engineers from advancing to permanent careers within the institution. It also opens the door to considering other established university policies that may hinder our country's ability to profit from creativity of all trained scientists, both male and female. For example, one policy that might be opened for reexamination is the usual requirement that all assistant professor appointments be full time. Part-time appointments would allow both women and men the opportunity to better balance family and career over time. This chapter presents the key findings from each of the preceding chapters, followed by recommendations and questions for future research.

KEY FINDINGS

As a foundation for understanding the survey findings, it is important to remember that **although women represent an increasing share of science, mathematics, and engineering faculty, they continue to be underrepresented in many of those disciplines.** While the percentage of women among faculty in scientific and engineering overall increased significantly from 1995 through

2003, the degree of representation varied substantially by discipline, and there remained disciplines where the percentage of women was significantly lower than the percentage of men. Table 6-1 shows the percentage of female faculty in selected scientific and engineering disciplines during this time period at the assistant, associate, and full professor levels.

In 2003, women comprised 20 percent of the full-time employed science and engineering (S&E) workforce and had slowly gained ground compared to men in the full-time academic workforce; by 2003, they represented about 25 percent of academics. Women's representation in the academic workforce, of course, varied by discipline: in the health sciences, women were the majority of full-time, employed doctorates, while in engineering they were less than 10 percent. The greatest concentration of women among full-time academics was at medical schools; the lowest was at Research II institutions.

Chapter 3—Academic Hiring

The findings on academic hiring suggest that many women fared well in the hiring process at Research I institutions, which contradicts some commonly held perceptions of research-intensive universities. If women applied for positions at RI institutions, they had a better chance of being interviewed and receiving offers than had male job candidates. Many departments at Research I institutions, both public and private, have made an effort to increase the numbers and percentage of female faculty in the sciences, engineering, and mathematics. Having women play a visible role in the hiring process, for example, has clearly made a difference. Unfortunately, women continue to be underrepresented in the applicant pool, relative to their representation among the pool of recent Ph.D.s. Institutions may not have effective recruitment plans, as departmental efforts targeted at women were not strong predictors in these surveys of an increased percentage of female applicants.

Applications

Finding 3-1: Women accounted for about 17 percent of applications for both tenure-track and tenured positions in the departments surveyed. There was wide variation by field and by department in the number and percentage of female applicants for faculty positions. In general, the higher the percentage of women in the Ph.D. pool, the higher the percentage of women applying for each position in that field, although the fields with lower percentages of women in the Ph.D. pool had a higher propensity for those women to apply (see Table 6-2). The percentage of applicant pools that included at least one woman was substantially higher than would be expected by chance. However, there were no female applicants (only men applied) for 32 (6 percent) of the available tenure-track positions and 16 (16.5 percent) of the tenured positions.

TABLE 6-1 Representation of Women in Faculty Positions at Research I Institutions by Rank and Field (percent), 1995-2003

	Assistant Professor					Associate Professor					Full Professor				
	1995	1997	1999	2001	2003	1995	1997	1999	2001	2003	1995	1997	1999	2001	2003
Agriculture	17.8	18.6	19.6	18.1	27.2	12.7	12.5	10.7	17.6	13.9	4.9	5.2	6.1	6.6	8.0
Biology	35.6	38.2	36.0	37.0	38.8	26.0	24.3	26.3	30.2	31.2	14.0	14.1	15.8	18.0	20.8
Engineering	14.2	12.7	12.8	14.8	16.6	4.8	6.4	9.6	9.3	11.7	1.8	1.4	2.3	2.7	3.8
Health sciences	69.1	66.9	64.0	64.7	66.5	65.6	65.1	64.9	64.5	59.1	35.1	38.9	45.3	48.0	59.0
Mathematics	18.7	22.0	26.5	25.2	26.6	10.4	14.4	14.9	15.8	16.3	7.6	5.9	9.9	10.0	9.7
Physics	25.1	25.6	24.6	25.4	24.1	9.5	13.4	14.8	16.7	19.5	4.3	4.6	5.9	6.8	7.6

SOURCE: National Science Foundation, Survey of Doctorate Recipients, 1995-2003. Tabulated by the National Research Council.

TABLE 6-2 Transitions from Ph.D. to Tenure-Track Positions by Field at the
Research I Institutions Surveyed (percent)

	Doctoral Pool	Pools for Tenure-Track Positions		
	Percent Women Ph.D.s (1999-2003)	Mean Percent of Applicants Who Are Women	Mean Percent of Applicants Invited to Interview Who Are Women	Mean Percent of Offers that Go to Women
Biology	45	26	28	34
Chemistry	32	18	25	29
Civil engineering	18	16	30	32
Electrical engineering	12	11	19	32
Mathematics	25	20	28	32
Physics	14	12	19	20

SOURCE: Survey of departments carried out by the Committee on Gender Differences in Careers of
Science, Engineering, and Mathematics Faculty; Ph.D. data is from the NSF, WebCASPAR.

**Finding 3-3: In each of the six disciplines, the percentage of applications from
women for tenure-track positions was lower than the percentage of Ph.D.s
awarded to women.**

Table 6-2 shows the percentage of women in the pool at each of several key
transition points in academic careers: award of Ph.D., application for position,
interview, and job offer. In each discipline, the percentage of applications from
women was lower than the percentage of doctoral degrees awarded to women.
This was particularly the case in chemistry and biology, the two disciplines in
the study with the highest percentage of female Ph.D.s. The mean percentage
of female applicants for tenure-track positions in chemistry was 18 percent, but
women earned 32 percent of the Ph.D.s in chemistry from Research I institutions
from 1999-2003. Biology (24 percent in the tenure-track pool and 45 percent in
the doctoral pool) also showed a significant difference. Electrical engineering (10
percent in the tenure-track pool and 12 percent in the doctoral pool), mathematics,
and physics had modest decreases in the applicant pool.

Recruitment

**Finding 3-7: Most of the institutional and departmental strategies that were
proposed for increasing the proportion of women in the applicant pool were
not strong predictors of the percentage of women applying. Most steps (such
as targeted advertising and recruiting at conferences) were done in isolation,
with almost two-thirds of the departments in our sample reporting that they
took either no steps or only one step to increase the gender diversity of the
applicant pool.**

Finding 3-8: **The percentage of women on the search committee and whether a woman chaired the committee were both significantly and positively associated with the percentage of women in the applicant pool (p = 0.01 and p = 0.02, respectively).**

Interviews

Finding 3-10: The percentage of women who were interviewed for tenure-track or tenured positions was higher than the percentage of women who applied. For each of the six disciplines in this study the mean percentage of females interviewed for tenure-track and tenured positions exceeded the mean percentage of female applicants. For example, the female applicant pool for tenure-track positions in electrical engineering was 11 percent, and the corresponding interview pool was 19 percent.

Finding 3-11: Although the percentage of women in interview pools across the six disciplines exceeded the percentage of women in applicant pools, no women were interviewed for 28 percent (155 positions) of the tenure track and 42 percent (42 positions) of the tenured jobs. These figures are substantially higher than those for men. However, the percentage of male applicants was much higher than the percentage of female applicants, and part of this number was comprised of cases for which there were no female applicants.

Job Offers

Finding 3-13: For all disciplines the percentage of tenure-track women who received the first job offer was greater than the percentage in the interview pool. Women received the first offer in 29 percent of the tenure-track and 31 percent of the tenured positions surveyed. Tenure-track women in all of these disciplines received a percentage of first offers that was greater than their percentage in the interview pool. For example, women were 21 percent of the interview pool for tenure-track electrical engineering positions and received 32 percent of the first offers. This finding is also true for tenured positions with the notable exception of biology, where the interview pool was 33 percent and women received 22 percent of the first offers.

Finding 3-14: In 95 percent of the tenure-track and 100 percent of the tenured positions where a man was the first choice for a position, a man was ultimately hired. In contrast, in cases where a woman was the first choice, a woman was ultimately hired in only 70 percent of the tenure-track and 77 percent of the tenured positions. When faculty were asked what factors they considered when selecting their current position, the effect of gender was statistically significant for only one factor—"family-related reasons."

Chapter 4—Professional Activities,
Institutional Resources, Climate, and Outcomes

The survey findings with regard to climate and resources demonstrate two critical points. First, discipline matters, as indicated by the difference in the amount of grant funding held by male and female faculty in biology, but not in other disciplines. Second, institutions have been doing well in addressing most of the aspects of climate that they can control, such as start-up packages and reduced teaching loads. Where the challenge may remain is in the climate at the departmental level. Interaction and collegial engagement with one's colleagues is an important part of scientific discovery and collaboration, and here female faculty were not as connected.

Professional Activities

Finding 4-1: There is little evidence overall that men and women spent different proportions of their time on teaching, research, and service. There is some indication that men spent a larger proportion of their time on research and fundraising than did women (42.1 percent for men compared to 40 percent for women). However, the difference only approaches significance, and the actual percentages of time that male and female faculty reported spending on research were not very different, with the exception of chemistry, for which men spent a significantly greater percentage of their time on research and fundraising (45.7 percent) than did women (39 percent) and mathematics (44.2 percent for men compared to 38.2 percent for women).

Finding 4-2: Male and female faculty appeared to have taught the same amount (41.4 percent for men compared to 42.6 percent for women). There were no gender differences in the number of undergraduate or graduate courses men and women taught: 0.83 undergraduate courses for men compared to 0.82 undergraduate courses for women. The percentages not teaching graduate courses were 50.8 percent for men and 54.9 percent for women.

Institutional Resources

Male and female faculty appeared to have similar access to many kinds of institutional resources, although there were some where male faculty seemed to have an advantage.

Finding 4-3: Men and women seem to have been treated equally when they were hired. The overall size of start-up packages and the specific resources of reduced initial teaching load, travel funds, and summer salary did not differ between male and female faculty.

Finding 4-4: Male and female faculty supervised about the same number of research assistants and postdocs.

Finding 4-5: There were some resources where male faculty appeared to have an advantage. These included the amount of laboratory space (considering both faculty overall and only those who do experimental research); access to equipment needed for research; and access to clerical support.

The apparent gender differences in access to these resources may reflect differences in access based on discipline or rank, since some disciplines and ranks have a higher proportion of male faculty, and those disciplines and ranks could also have more lab space and equipment.

Climate

Professional climate may be somewhat different for male and female faculty.

Finding 4-6: Female tenure-track and tenured faculty reported that they were more likely to have mentors than male faculty. In the case of tenure-track faculty, 57 percent of women had mentors compared to 49 percent of men.

Finding 4-7: Female faculty reported that they were less likely to engage in conversation with their colleagues on a wide range of professional topics. These topics included research, salary, and benefits (and, to some extent, interaction with other faculty members and departmental climate). This distance may prevent women from accessing important information and may make them feel less included and more marginalized in their professional lives. Male and female faculty did not differ in their reports of discussions with colleagues on teaching, funding, interaction with administration, and personal life.

Finding 4-8: There were no differences between male and female faculty on two measures of inclusion: chairing committees (39 percent for men and 34 percent for women) and being part of a research team (62 percent for men and 65 percent for women).

Outcomes

There is little evidence across the six disciplines that men and women have exhibited different outcomes on most key measures (including publications, grant funding, nominations for international and national honors and awards, salary, and offers of positions in other institutions). On all measures, there were significant differences among disciplines.

Finding 4-9: Overall, male faculty had published marginally more refereed articles and papers in the past 3 years than female faculty, except in electrical engineering, where the reverse was true. Men had published significantly more papers than women in chemistry (men, 15.8; women, 9.4) and mathematics (men, 12.4; women, 10.4). In electrical engineering, women had published marginally more papers than men (women, 7.5; men, 5.8). The differences in the numbers of publications between men and women were not significant in biology, civil engineering, and physics. All the other variables related to the number of published articles and papers (discipline, rank, prestige of institution, access to mentors, and time on research) show the same effects for male and female faculty.

Finding 4-10: Although men were somewhat less likely to be a principal investigator or co-principal investigator on a grant proposal than were women, this difference disappeared when other variables were added in a regression analysis, where male and female faculty did not differ on the probability of having grant funding. Furthermore, because the effect of gender was confounded with the effect of rank and whether the person had a mentor, it is essentially impossible to isolate the effect of gender. The variables that appear to be associated with the probability of having a grant (discipline, faculty rank, being at a high- or medium-prestige university, and spending more time on research) do so in the same way for male and female faculty.

Finding 4-11: Male faculty had significantly more research funding than female faculty in biology; in the other disciplines, the differences between male and female faculty were not significant. There was no overall difference in the amount of grant funding received by male and female faculty, but there was a significant interaction between gender and discipline. The other variables related to the amount of grant funding (faculty rank, whether a faculty member is at a private university, whether a faculty member is at a university of higher prestige, having a mentor, and publishing more) were related in the same way for male and female faculty.

Finding 4-12: Female assistant professors who had a mentor had a higher probability of receiving grants than those who did not have a mentor. In chemistry, female assistant professors with mentors had a 95 percent probability of having grant funding compared to 77 percent for female assistant professors in chemistry without mentors. A similar but weaker pattern is exhibited for female associate professors. Over all six fields surveyed female assistant professors with no mentors had a 68 percent probability of having grant funding compared to 93 percent of women with mentors. This contrasts with the pattern for male assistant professors; those with no mentor had an 86 percent probability of having grant funding compared to 83 percent for those with mentors.

Finding 4-13: Overall male and female faculty were equally likely to be nominated for international and national honors and awards, but the results varied significantly by discipline, making interpretation challenging. The other variables affecting the likelihood of being nominated for honors and awards (discipline, faculty rank, prestige of university, number of publications) affected this likelihood in the same way for male and female faculty.

Finding 4-14: Gender was a significant determinant of salary, but only among full professors. Male full professors made, on average, about 8 percent more than women, once we controlled for discipline. At the associate and assistant professor ranks, the differences in salaries of men and women disappeared.

Finding 4-15: Differences in the probability of receiving an outside offer for male and female faculty depended on discipline. In electrical engineering and in mathematics women were more likely to have received an outside offer, while the trend was reversed in chemistry and physics.

Chapter 5—Tenure and Promotion

The findings related to tenure and promotion indicate the importance of addressing the retention of women faculty in the early stages of their academy careers; not as many were considered for tenure as would be expected, based on the number of female assistant professors. Retention was particularly problematic given the increased duration of time in rank for all faculty. Both male and female faculty utilized stop-the-tenure-clock policies—spending a longer time in the uncertainty of securing tenure—but women used these policies more. Female faculty who did come up for tenure were as successful or more successful than men, so one of the most important challenges may be increasing the pool of female faculty who make it to that point.

Award of Tenure

Finding 5-1: In every field, women were underrepresented among candidates for tenure relative to the number of female assistant professors. Most strikingly, women were most likely to be underrepresented in the fields in which they accounted for the largest share of the faculty—biology and chemistry. In biology and chemistry, the differences were statistically significant. In biology, 27 percent of the faculty considered for tenure were women, although women represented 36 percent of the assistant professor pool. In chemistry those numbers were 15 percent and 22 percent, respectively. This difference may suggest that female assistant professors were more likely to leave before being considered for tenure than were men. It might also reflect increased hiring of female assistant professors in recent years (compared with hiring 6 to 8 years ago).

Finding 5-2: Given that the interaction between the gender of the candidate and the percentage of women in the tenure-track pool was statistically significant ($p = 0.012$), women appeared to be more likely to be promoted when there was a smaller percentage of women among the tenure-track faculty, resulting in a greater difference between men and women in their tenure success in departments with fewer female assistant professors.

Finding 5-3: Women were more likely than men to receive tenure when they came up for tenure review. When controlling only for field and gender of the candidate, we found that women were marginally more likely than men to receive tenure ($p = .0567$). Women received tenure in 92 percent of the cases (115 out of 125) compared to 87 percent of the cases for men (548 out of 633).

Finding 5-4: Discipline, stop-the-tenure-clock policies, and departmental size were not associated with the probability of a positive tenure decision for either male or female faculty members who were considered for tenure. Both male and female assistant professors were significantly more likely to receive tenure at public institutions (92 percent) compared to private institutions (85 percent; $p = 0.029$).

Finding 5-5: Eighty-eight percent of both male and female survey respondents stated that they knew their institution's policy on tenure. Eighty-one percent of male faculty knew their institution's policies on promotion. However, only 75 percent of female faculty respondents knew their institution's policy on promotion, which is statistically significant ($p = 0.02$).

Promotion to Full Professor

No significant gender disparity was found at the stage of promotion to full professor.

Finding 5-6: For the six disciplines surveyed, 90 percent of the men and 88 percent of the women proposed for full professor were promoted—a difference that was not statistically significant. There was no significant difference in the probability of promotion to full professor due to gender of the candidate, after accounting for other potentially important factors such as disciplinary differences, departmental size, and use of stop-the-tenure-clock policies. Once proposed for promotion to full professor, women and men appeared to have fared about the same across all types of institutions and departments.

Finding 5-7: Women were proposed for promotion to full professor at approximately the same rates as they were represented among associate professors. Female faculty in biology were considered for promotion in 24 percent of the cases (28 percent of the associate professor pool); 14 percent of the cases in chemistry (18 percent of the pool); 18 percent of the cases in civil engineering (14 percent of the pool); 17 percent of cases in electrical engineering (13 percent of the pool); 9 percent of cases in mathematics (15 percent of the pool); and 7 percent of the cases in physics (8 percent of the pool).

Time in Rank

Women spent significantly longer time in rank as assistant professors than men did.

Finding 5-8: Time in rank as an assistant professor has grown over time for both male and female faculty. Men who were full professors at the time of the survey had spent the least amount of time in rank as assistant professors. This was true across all disciplines.

Finding 5-9: Women who were associate professors at the time of the survey had averaged a significantly longer time in rank as assistant professors in all fields except electrical engineering, where women's shorter time in rank was not significantly different. It is difficult to determine whether these apparent differences persist once we control for individual and departmental characteristics such as length of postdoctoral experience and stopping the tenure clock for family leave. While women did appear to remain at the rank of assistant professor longer than men, the differences between genders depended upon factors such as the prestige of the institution, the time elapsed since the completion of the doctoral degree, and the current rank of the individual. Both male and female faculty spent longer time in the assistant professor ranks at institutions of higher prestige.

Finding 5-10: Male and female faculty who stopped the tenure clock spent significantly more time as assistant professors than those who did not (an average of 74 months compared to 57 months). They had a lower chance of promotion to associate professor (about 80 percent) at any time (given that they had not been promoted until then) than those who did not stop the clock. Everything else being equal, however, stopping the tenure clock did not affect the probability of promotion and tenure; it just delayed it by about a year and a half. It is unclear how that delay affected female faculty, who were more likely than men to avail themselves of this policy. Although the effect of stopping the tenure clock on the probability of promotion and tenure was similar for both male and female faculty, 19.7 percent of female assistant professors in the survey sample availed themselves of this policy compared to 7.4 percent of male assistant professors. At the associate professor level, 10.2 percent of female faculty compared to 6.4 percent of male faculty stopped the tenure clock.

Time from Receipt of Ph.D.

Finding 5-12: Overall, it appears that female faculty took significantly longer from receipt of Ph.D. to promotion to associate professor with tenure, but this gender effect was confounded with current rank, discipline, and other factors. It is difficult to determine whether these apparent differences persist once we control for individual and departmental characteristics such as length of postdoctoral experience and stopping the tenure clock for family leave. While women did appear to remain at the rank of assistant professor longer than did men, the differences between gender depended on factors including the prestige of the institution, the time elapsed since completion of the doctoral degree, and the current rank of the individual.

RECOMMENDATIONS

The survey data suggest that positive changes have taken place and continue to occur. At the same time, the data should not be mistakenly interpreted as indicating that male and female faculty in math, science, and engineering have reached full equality and representation, and we caution against premature complacency. Women remain underrepresented among science and engineering faculty and in the tenure-track applicant pool for faculty positions in all disciplinary areas examined. Furthermore, few departments surveyed reported extensive efforts to increase gender diversity of the applicant pool. Much work remains to be done by institutions and professional disciplinary societies to accomplish full representation of men and women in academic departments. And much additional research is needed to understand the full career paths of female academics, from receipt of Ph.D. to retirement, and to document gender differences in other disciplines, other types of institutions, and other types of faculty positions.

Recommendations for Institutions

Research I institutions should:

1. Design and implement new programs and policies to increase the number of women applying for tenure-track or tenured positions and evaluate existing programs for effectiveness. This includes enhancing institutional efforts to encourage female graduates and postdocs to consider careers at RI institutions. In each of the six disciplines studied, women were underrepresented in the applicant pool relative to their representation in the pool of recent Ph.D.s (Finding 3-3). This critical gap must be narrowed to expand the number of female faculty in research-intensive institutions. Most departments reported using a very small arsenal of recruitment strategies (targeted advertising was the most cited), and 43 percent reported using only one strategy (see Finding 3-7). Significant change in the applicant pool will not come from such minimal efforts.

2. Involve current female faculty in faculty searches, with appropriate release time. The proportion of women on the search committee and whether a woman chaired the committee were both significantly and positively associated with the proportion of women in the applicant pool (see Finding 3-8). Such engagement may signal to prospective hires that the institutional climate is supportive and inclusive.

3. Investigate why female faculty, compared to their male counterparts, appear to continue to experience some sense of isolation in subtle and intangible ways. Finding 4-7, for example, reports that female faculty are less likely to engage with other faculty in conversations about research or salary. Creating informal opportunities for faculty to engage within a department or across an institution might help to address this issue.

4. Explore gender differences in the obligations outside of professional responsibilities (particularly family-related obligations) and how these differences may affect the professional outcomes of their faculty. Our findings focused only on the climate within academic institutions, but factors outside the institutional environment may be equally important. (Findings 4-6 through 4-8).

5. Initiate mentoring programs for all newly hired faculty, especially at the assistant professor level. As described in Finding 4-12, the mentoring of female faculty had a striking impact on their ability to secure grant funding. Institutional mentoring programs could help to ensure that female faculty acquire grant funding, which in turn should have a positive effect on their promotion rates.

6. Make tenure and promotion procedures as transparent as possible and ensure that policies are routinely and effectively communicated to all faculty. While 81 percent of male faculty know their institution's policies on promotion, only 75 percent of female faculty do (see Finding 5-5). Departments in particular need to review their communication strategies, as only 49 percent of all faculty surveyed reported that their department had written procedures. And only 78 percent of departments reported that they had written tenure and promotion policies.

7. Monitor and evaluate stop-the-tenure-clock policies and their impact on faculty retention and advancement. Where such policies are not already in place, adopt them and ensure effective dissemination to faculty members. Only 78 percent of assistant professors reported that their department or university had a formal family or personal leave policy that allows stopping or extending the tenure clock. At those institutions that do, 19.7 percent of female and 7.4 percent of male assistant professors avail themselves of these policies, as well as 10.2 percent of female and 6.4 percent of male associate professors (see Finding 5-10). As use of these policies will likely grow, institutions need to review the careers of faculty who use these policies to understand their impact on career progress.

8. Collect data encompassed in this study (including applications, interviews, first offers, hires, time in rank, tenure award, and promotion) disaggregated by race, ethnicity, and gender. Many of the departments surveyed have made significant gains in their numbers of female faculty at many of these critical junctures, yet these results are not well known. The collection of data can allow departments and institutions to focus their scarce resources on transitions that need the most attention. Also, our findings do not address race and ethnicity, but this information is essential as institutions work to increase diversity.

Recommendations for Professional Societies

Professional societies in science and engineering disciplines should:

9. Collect data on the career tracks of their members. This study identified many differences among disciplines that warrant investigation. Why, for example, do biology and chemistry have disproportionately smaller applicant pools of women for faculty positions? (Finding 3-3) And why are women in electrical engineering and mathematics more likely than men to receive outside job offers, while the reverse is true for chemistry and physics? (Finding 4-15)

10. Disseminate successful strategies to increase the gender diversity of the applicant pools for tenure-track and tenured faculty positions. Only 10 percent of departments reported relying on three or more strategies for recruitment. (Table 3-10)

11. Conduct in-depth surveys of their members at regular intervals on the climate for professional success and the role of mentoring in their discipline. (Findings 4-6, 4-7, and 4-12)

Questions for Future Research

This study raises many unanswered questions about the status of women in academia. As noted at the onset of this report, the surveys did not capture the experiences of Ph.D.s who never apply for academic positions, nor of female faculty who have left at various points in their academic careers. We also recognize that there are important, nonacademic issues affecting men and women differently that impact career choices at critical junctures. Fuller examination of these issues (for example, topics relating to family, children, home life, care of elderly parents) will shed greater light on career choices by women and men and should yield suggestions on the types of support needed to encourage retention of women in academic careers. Below are suggestions for future research:

A Deeper Understanding of Career Paths

1. Using longitudinal data, what are the academic career paths of women in different science and engineering disciplines from receipt of their Ph.D. to retirement? Most importantly, where do women Ph.D.s go who do not apply for academic positions, and where do women faculty go who leave the university before tenure consideration?

2. Why are women underrepresented in the applicant pools and among those who are considered for tenure? How can we understand more fully the subtle but powerful influences of climate and family life on career decisions? While it is true that the lives of female faculty have become more similar to those of men in recent years, the discrepancies remain very large, which may be a major reason why women don't consider careers in RI institutions. The demands of family life are also a large deterrent. Universities can do a lot by mentoring of female graduate students that it is possible to have a career at an RI institution and still have a family life.

3. Why aren't more women in fields such as biology and chemistry applying to RI tenure-track positions, as discussed in Finding 3-3? Such a study might examine the career preferences of graduate students and postdocs (and what factors shape those preferences) as well as the efforts of departments and institutions to recruit faculty in these disciplines.

4. Why do female faculty, compared to their male counterparts, appear to continue to experience some sense of isolation in more subtle and intangible areas? The findings on institutional climate indicate several areas that still need to be examined to facilitate the full participation of all faculty. Finding 4-7, for example, reports that female faculty are less likely to engage with other faculty in conversations about research or salary.

5. What is the impact of stop-the-tenure-clock policies on faculty careers? Given the significant increases in the number of faculty invoking stop-the-tenure-clock policies there is a need to collect longitudinal data on the career patterns of these faculty including data on time in rank, tenure, and promotion statistics. Does this extension of uncertainty regarding tenure for assistant professors who utilize their institutions' stop-the-tenure-clock policies deter a certain fraction of women (and men) from applying or have a negative effect on the promotion and retention of faculty who utilize these policies?

6. What are the causes for the attrition of women and men prior to tenure decisions, if indeed attrition does take place? This is particularly relevant given Finding 5–9, which indicates that female faculty spend significantly longer in time in rank as assistant professors, and this may have an impact on retention of female faculty.

7. To what extent are female faculty rewarded beyond promotion to full professor? There are career milestones beyond promotion to full professor in academia. A future study that looks at chaired professorships, salary increments, and continued access to institutional resources would be useful.

8. What important, nonacademic issues affect men and women differently that impact their career choices at critical junctures? While the committee was not able to investigate them in this study, a fuller examination—for example, of issues relating to family, children, home life, care of elderly parents, etc.—might shed light on career choices by men and women and offer suggestions on the nature and types of supports to encourage retention of women pursuing academic careers in science, engineering, and mathematics.

Expanding the Scope

9. How important are differences among fields? Future studies should examine additional engineering and scientific fields because as the data in this report demonstrates fields differ a lot from each other. Certain engineering fields, including chemical engineering and bioengineering, may look very different from the two engineering fields—civil and electrical—examined here.

10. What are the experiences of faculty at Research II institutions? There would be value in expanding the scope of this study. Conduct further research to understand the hiring efforts and results at Research II universities (which also conduct research and train doctorates). Past research suggests that female faculty in science and engineering are the least well-represented at Research II institutions, with an average percentage of 15 percent.

11. What are the experiences of part-time and non-tenure track faculty? A significant but necessary limitation of this study is that it focused on full-time tenure-track and tenured faculty. Given that the population of non-tenure track and part-time faculty is growing, and that a good portion of these faculty are women, it would be very valuable to have data and information on the careers of these faculty.

Appendixes

Appendix 1-1
Biographical Information on Committee Members

Claude R. Canizares (Co-Chair) is the Vice President for Research and Associate Provost and the Bruno Rossi Professor of Physics at the Massachusetts Institute of Technology (MIT). He has overall responsibility for research activity and policy at MIT, overseeing more than a dozen interdisciplinary research laboratories and centers including the MIT Lincoln Laboratory, the Broad Institute, the Plasma Science and Fusion Center, the Research Laboratory of Electronics, the Institute for Soldier Nanotechnology, the Francis Bitter Magnet Laboratory, Haystack Observatory, and the Division of Health Sciences and Technology. He oversees several offices dealing with research policy and administration; he chairs the Research Policy Committee and serves on the Academic Council and the Academic Appointments committee among others. He serves on the National Research Council (NRC) committees on Science Engineering and Public Policy and Science Communication and National Security, and he has served on the Council of the National Academy of Sciences and as chair of the Space Studies Board. He is on the Board of Directors of L-3 Communications, Inc. Professor Canizares is a member of the National Academy of Sciences and the International Academy of Astronautics and is a fellow of the American Academy of Arts & Sciences, the American Physical Society, and the American Association for the Advancement of Science. Professor Canizares is the Associate Director of the Chandra X-ray Observatory Center and a principal investigator on NASA's Chandra X-ray Observatory, having led the development of the Chandra High Resolution Transmission Grating Spectrometer. His main research interests are high resolution x-ray spectroscopy and plasma diagnostics of supernova remnants and clusters of galaxies, X-ray studies of dark matter, X-ray properties of quasars and active galactic nuclei, and observational cosmology. He is author or co-author of more than 200 scientific papers. Professor Canizares earned his B.A., M.A., and Ph.D. in physics from Harvard University. He went to MIT as a postdoctoral fellow in the Physics Department in 1971 and joined the faculty in 1974. Professor Canizares has received several awards including decoration for Meritorious Civilian Service to the United States Air Force, two NASA Public Service Medals, and the Goddard Medal of the American Astronautical Society.

Sally E. Shaywitz (Co-Chair) is the Audrey G. Ratner Professor in Learning Development at the Yale University School of Medicine and Co-Director of the Yale Center for Learning, Reading and Attention, and the newly formed Yale Center for Dyslexia & Creativity. Dr. Shaywitz, an elected member of the Institute of Medicine (IOM) of the National Academies, has served as Chair of her Section and on the Membership Committee of the IOM. In recognition of her scientific contributions, she was awarded an Honorary Doctor of Science degree from Williams College; the Townsend Harris Medal of the City College of New York; the Annie Glenn Award for Leadership from the Ohio State University; the Achievement Award in Women's Health of the Society for the Advancement of Women's Health Research; and the Distinguished Alumnus Award of the Albert Einstein College of Medicine. In recognition of her contributions to the National Academy of Sciences, Dr. Shaywitz was named a National Associate of the National Academies. Dr. Shaywitz served on the Advisory Council of the National Institute of Neurological Diseases and Stroke (NINDS), on the National Research Council Committee on Women in Science and Engineering, and the Scientific Advisory Board of the March of Dimes; she currently serves on the National Advisory Board of Recordings for the Blind and Dyslexic and on the National Board of the Institute for Educational Sciences of the Department of Education. Dr. Shaywitz currently co-chairs the National Research Council Committee on Gender Differences in the Careers of Science, Engineering, and Mathematics Faculty; she most recently presented at the Gordon Research Conference on the Auditory Cortex and served on the Institute of Medicine Committee on Understanding the Biology of Sex and Gender Differences and the National Reading Panel and the Committee to Prevent Reading Difficulties in Young Children of the National Research Council. Dr. Shaywitz is the author of more than 200 scientific articles, chapters, and books, including *Overcoming Dyslexia* (Knopf, 2003). Her research provides the basic framework: conceptual model, epidemiology and neurobiology for the scientific study of dyslexia. Dr. Shaywitz originated and championed the "Sea of Strengths" model of dyslexia, which emphasizes a sea of strengths of higher critical thinking and creativity surrounding the encapsulated weakness found in children and adults who are dyslexic. Dr. Shaywitz received her A.B. (with Honors) from the City University and her M.D. from Albert Einstein College of Medicine.

Linda Abriola is Dean of Engineering at Tufts University. Prior to that appointment, she was the Horace Williams King Collegiate Professor of Environmental Engineering at the University of Michigan. Dr. Abriola received Ph.D. and master's degrees in civil engineering from Princeton University and a bachelor's degree in civil engineering from Drexel University. Her primary research focus is the integration of mathematical modeling and laboratory experiments to investigate and elucidate processes governing the transport, fate, and remediation of nonaqueous phase liquid organic contaminants in the subsurface. Dr. Abriola's numerous professional activities have included service on the U.S. Environmental Protection Agency Science Advisory Board, the National Research Council Water Science and Technology Board, and the U.S. Department of Energy's NABIR (Natural and Accelerated BIoremediation Research) Advisory Committee. An author of more than 130 refereed publications, Dr. Abriola has been the recipient of a number of awards, including the Association for Women Geoscientist's Outstanding Educator Award (1996), the

National Ground Water Association's Distinguished Darcy Lectureship (1996), and designation as a ISI Highly Cited Author in Ecology/Environment (2002). She is a fellow of the American Geophysical Union and a member of both the American Academy of Arts & Sciences and the National Academy of Engineering (NAE). Dr. Abriola is an elected member of the American Society of Engineering Education Engineering Dean's Council and the NAE governing council.

Jane Buikstra is a bioarchaeologist and is a Regents' Professor at Arizona State University where she also directs the Center for Bioarchaeological Research within the School of Human Evolution and Social Change. She was formerly the Leslie Spier Distinguished Professor of Anthropology at the University of New Mexico, and the Harold H. Swift Distinguished Service Professor of Anthropology at the University of Chicago. She received her Ph.D. from the University of Chicago. She was elected to the National Academy of Sciences in 1987. Her research interests include paleopathology, human skeletal biology, paleodemography, forensic anthropology, genetic relationships within and between paleopopulations, paleodiet, and funerary archaeology. She teaches human osteology, paleopathology, bioarchaeology, forensic anthropology, archaeology of death, and field archaeology. She is a past President of the American Anthropological Association, the American Association of Physical Anthropologists, and the Paleopathology Association and is currently President of the Center for American Archeology. She has received numerous research grants from the National Science Foundation, the National Endowment for the Humanities, the Wenner-Gren Foundation, the National Geographic Society, and the Smithsonian Institution. She has authored or edited 19 books, including the *Bioarchaeology of Tuberculosis: A Global View of a Re-Emerging Disease* (2003) with Charlotte Roberts, and *Bioarchaeology: The Contextual Approach to the Study of Human Remains* (2006) with Lane Beck. In addition, she has published more than 100 articles or chapters on a variety of subjects, including bone chemistry in eastern North America, ancient treponematosis and tuberculosis in the Americas and in Egypt, diet and health of Argaric peoples (Bronze Age, Spain), australopithecine spinal pathology, trauma in Copan's founding dynasty (Maya), coca-chewing, cranial deformation, tuberculosis, and funerary rituals of ancient Andeans.

Alicia Carriquiry is Professor of Statistics and has served as Associate Provost, Iowa State University. Dr. Carriquiry is an elected member of the International Statistical Institute, a fellow of the American Statistical Association, and a fellow of the Institute of Mathematical Statistics. She is Past President of the International Society for Bayesian Analysis, and served on the Executive Committee of the Institute of Mathematical Statistics. She has been a trustee of the National Institute of Statistical Sciences since 1997, and has served on its Executive Committee. She is currently Vice President of the American Statistical Association and a member of the Council of the International Statistical Institute. Dr. Carriquiry is Associate Editor of *Statistical Sciences* and Associate Editor of *The Annals of Applied Statistics* and serves on the editorial boards of two Latin American journals in mathematics and statistics. She currently serves on the Human Subjects Review Board of the U.S. Environmental Protection Agency and is a consultant to the Chilean government on the upcoming National Health Survey. Dr. Carriquiry has published more than 90 refereed articles and technical reports and has co-edited four books. Her research interests are in the development of Bayesian methods and applications in public health, nutrition, traffic safety, and genetics. She has also collaborated in research projects in the area of stochastic volatility and other nonlinear models for time-dependent data. She has served on several National Academy of Sciences committees, in addition to the present one. She participated in the Subcommittee on Interpretation and Uses of the Dietary Reference Intakes; the standing Committee on Applied and Theoretical Statistics; the Committee on Assessing the Feasibility and Technical Capabilities of a National Ballistics Database; the Committee on Eligibility for the Women and Infant Children Program; and the Committee on Ethics and Scientific Validity of Toxicity Studies Involving Human Subjects. She currently serves on the standing Committee on National Statistics and on the standing Committee on the Use of Evidence in Public Policy. She was recently a reviewer of *Beyond Bias and Barriers.*

Ronald G. Ehrenberg is the Irving M. Ives Professor of Industrial and Labor Relations and Economics and a Stephen H. Weiss Presidential Fellow at Cornell University, as well as the Director of the Cornell Higher Education Research Institute. He currently is an elected member of Cornell's Board of Trustees and from July 1, 1995, to June 30, 1998, served as Cornell's Vice President for Academic Programs, Planning and Budgeting. Dr. Ehrenberg received his B.A. in mathematics from Harpur College (Binghamton University) in 1966, a Ph.D. in economics from Northwestern University in 1970, and an Honorary Doctor of Science from the State University of New York in 2008. A member of the Cornell faculty for 33 years, Dr. Ehrenberg has authored or co-authored more than 120 papers and authored or edited 20 books. He is a research associate at the National Bureau of Economic Research, a past President and fellow of the Society of Labor Economists, a fellow of the TIAA-CREF Institute, a fellow of the American Educational Research Association, a member of the National Academy of Education, and a member of the National Academy of Social Insurance. Dr. Ehrenberg previously chaired the American Association of University Professors committees on the economic status of the profession and on retirement, and served on the American Economic Association's Committee on the Status of Women in the Economics Profession. At the NRC he previously served on the Committee on Dimensions, Causes and Implications of Trends in Early Career Events for Life Scientists, the Committee on Methods for Forecasting Demand and Supply of Doctoral Scientists and Engineers, the Policy and Gobal Affairs Oversight Committee, and the Office of Scientific and Engineering Personnel Advisory Board. He previously chaired the NRC's Board on Higher Education and the Workforce and is a National Associate of the National Academies.

Joan Girgus is Professor of Psychology and Special Assistant to the Dean of the Faculty at Princeton University. She has also served as Chair of the Psychology Department and Dean of the College at Princeton. Prior to going to Princeton, she served as a faculty member and dean at the City College of City University of New York (CUNY). Dr. Girgus has done research and written books and papers on perception and perceptual development, personality development, the transition from childhood to adolescence, and the psychosocial basis of depression. She has also written papers on undergraduate science education and on women in science. Her research has been supported by the National Science Foundation, the National Institute of Mental Health, the National Institute of Child Health and Human Development, the Ford Foundation, and CUNY. Dr. Girgus is one of the principals of The Learning Alliance, the first just-in-time provider of strategic expertise to college and university leaders. From 1993-2003, she was a member of the executive committee of the Pew Higher Education Roundtable and its successor, the Knight Higher Education Roundtable, which worked with a broad range of colleges and universities to identify "best practices" for academic restructuring, and was a consulting editor of *Policy Perspectives*, which published essays on major issues in higher education. From 1987-1999, she directed the Pew Science Program, a national program to improve undergraduate science education sponsored by the Pew Charitable Trusts. Dr. Girgus is currently a trustee of Adelphi University, the Wenner-Gren Foundation, and McCarter Theatre. She has also served on the Board of Trustees of the American Association on Higher Education (AAHE) and Sarah Lawrence College. Dr. Girgus received her B.A. from Sarah Lawrence College and both her M.A. and Ph.D. from the Graduate Faculty of the New School for Social Research in New York City.

Arleen Leibowitz is Professor of Public Policy at the University of California at Los Angeles (UCLA) School of Public Affairs. Dr. Leibowitz has conducted research in health and labor economics since obtaining her Ph.D. in economics at Columbia University. Dr. Leibowitz's research centers on investments in human capital and in health. She has examined the role of maternal education in investments in children, educational outcomes for children, the demand for child care, the effect of education on women's labor force participation, secular trends in women's labor supply, and the effect of maternity leave on new mothers' return to work. As a member of the Health Insurance Experiment team at RAND, she worked extensively in health economics and policy, studying the effect of cost-sharing on medical care expenditures, children's health care use, birth rates, expenditures for prescription and over-the-counter drugs, and managed care. Dr. Leibowitz led the economics team of the HIV Cost and Services Utilization Study. Currently, Dr. Leibowitz heads the Policy Core of the UCLA Center for HIV

Identification, Prevention, and Treatment Services, where her research examines how public policies, such as Medicaid and AIDS Drug Assistance Program, and private policies, such as managed care, affect the amount and kind of medical care obtained by persons living with HIV. Dr. Leibowitz is one of the core participants in the Blue Sky Group, which seeks to redirect the discussion of health care reform from an exclusive focus on incremental improvements in medical care and in insurance to a more comprehensive vision of the health system. Dr. Leibowitz served on the Committee on National Statistics of the NRC (CNSTAT) from 2001 to 2004.

Thomas N. Taylor is Distinguished Professor in the Department of Ecology and Evolutionary Biology and Biodiversity Reearch Institute at the University of Kansas. He is also Senior Curator of the Natural History Museum and Biodiversity Research Center, and Courtesy Professor for the Department of Geology. He has served as Director of the State of Kansas NSF Experimental Program to Stimulate Competitive Research (EPSCoR) Program. Dr. Taylor holds a B.A. degree in botany from Miami University (Oxford) and a Ph.D. in botany from the University of Illinois (1964). He was a postdoctoral fellow at Yale University. He has served on numerous state, national, and international committees including the National Science Foundation—Education and Human Resources Advisory Committee, National Science Foundation—GPRA-Performance Assessment Advisory Committee, National Science Foundation—MPSAC/EHRAC Committee to Review Undergraduate Education in the Math and Physical Sciences, Chair of the Strategic Planning and Assessment Committee for NIH BRIN KU Medical Center, Senator Pat Robert's Advisory Committee in Science, Technology, Future Kansas Implementation Advisory Committee, and Bioinformatics Core Advisory Committee. He served on the Polar Research Board for the NRC and as Faculty Advisor to the Chancellor of the Ohio Board of Regents, and on the Government-University-Industry Research Roundtable for the State of Ohio. Dr. Taylor has published more than 380 peer-reviewed research papers and authored more than eight edited books on various aspects of the paleobiology of Antarctic fossil biotas, biology and evolution of fossil microbes, and evolution of early land plants. Dr. Taylor has received numerous honors including the Research Achievement Award in the State of Kansas, Distinguished Scholar Award and Teaching Award from Ohio State University, and the Merit Award from the Botanical Society of America. Dr. Taylor is a member of the National Academy of Sciences and is currently a member of the National Science Board where he serves on the Committee on Education and Human Resources, Subcommittee on Polar Issues, and Committee on Strategy and Budget.

Lilian Shiao-Yen Wu is Program Executive, Global University Relations, IBM Technology Strategy and Innovation and a research scientist. In this position she manages IBM's portfolio of investments in projects to support research collaborations between IBM and universities. These research collaborations often include governments, foundations, or companies. Prior to this position she was Consultant for Corporate Technology Strategy Development and for most of her career a research scientist in applied mathematics at the IBM T.J. Watson Research Center in Yorktown Heights, New York. Her major research interests are mathematical modeling and risk analysis, particularly for the services industry and the electric and energy industries. She holds a B.S. in mathematics from the University of Maryland, College Park and M.S. and Ph.D. degrees from Cornell University. Her professional services include Chair of the National Research Council's Committee on Women in Science, Engineering, and Medicine; member of the S&E Workforce Committee of the Government-University-Industry Research Roundtable of the National Research Council; and member of NSF's Advisory Committee on International Science and Engineering and NSF's Corporate Alliance. She was a member of President Clinton's Committee of Advisors on Science and Technology (PCAST), and NSF's Committee on Equal Opportunity in Science and Engineering, and she served on the Advisory Committee of NSF's Engineering Directorate. Among her other professional services, she served on American Association for the Advancement of Science's Committee on Public Understanding of Science and Technology and Department of Energy's Secretary of Energy's Laboratory Operations Advisory Board. She received her Ph.D. in applied mathematics from Cornell University. Her major research interests are analysis and modeling of technology-enabled and people-intensive complex systems, particularly in the services sector. She is also a member of the Board of Trustees of the New School University, the President's Council of Olin College, and the Global Advisory Board of Fordham University School of Business.

Appendix 1-2
List of Research I Institutions

Arizona State University; Boston University; Brown University; California Institute of Technology; Carnegie-Mellon University; Case Western Reserve University; Colorado State University; Columbia University; Cornell University; Duke University; Emory University; Florida State University; Georgetown University; Georgia Institute of Technology; Harvard University; Howard University; Indiana University at Bloomington; Iowa State University; Johns Hopkins University; Louisiana State University; Massachusetts Institute of Technology; Michigan State University; New Mexico State University; New York University; North Carolina State University; Northwestern University; Ohio State University; Oregon State University; Pennsylvania State University; Princeton University; Purdue University; Rockefeller University; Rutgers, the State University of NJ; Stanford University; State University of New York at Buffalo; State University of New York at Stony Brook; Temple University; Texas A&M University; Tufts University; Tulane University; University of Alabama at Birmingham; University of Arizona; University of California-Berkeley; University of California-Davis; University of California-Irvine; University of California-Los Angeles; University of California-San Diego; University of California-San Francisco; University of California-Santa Barbara; University of Chicago; University of Cincinnati; University of Colorado; University of Connecticut; University of Florida; University of Georgia; University of Hawaii at Manoa; University of Illinois at Chicago; University of Illinois at Urbana-Champaign; University of Iowa; University of Kansas; University of Kentucky; University of Maryland at College Park; University of Massachusetts at Amherst; University of Miami; University of Michigan; University of Minnesota, Twin Cities; University of Missouri-Columbia; University of Nebraska-Lincoln; University of New Mexico; University of North Carolina at Chapel Hill; University of Pennsylvania; University of Pittsburgh; University of Rochester; University of Southern California; University of Tennessee, Knoxville; University of Texas at Austin; University of Utah; University of Virginia; University of Washington; University of Wisconsin-Madison; Utah State University; Vanderbilt University; Virginia Commonwealth University; Virginia Polytechnic Inst. & State Univ.; Washington University; Wayne State University; West Virginia University; Yale University; Yeshiva University.

Appendix 1-3
Committee Meeting Agenda

Committee on Gender Differences in Careers of Science, Engineering, and Mathematics Faculty

First Committee Meeting Agenda
First Committee Meeting
The National Academies Keck Center Rm. 204
Washington, DC
January 29-30, 2004
AGENDA

January 29, 2004

CLOSED SESSION

8:00-8:30	Continental Breakfast	
8:30-8:45	Welcome	**TAB 1**
	Claude R. Canizares, Committee Chair	
	Richard Bissell, Executive Director, PGA	
	Connie Citro, Acting Chief of Staff, CNSTAT	
	Jong-on Hahm, Study Director and Director, CWSE	
8:45-9:15	Bias Discussion	
	Richard Bissell, Executive Director, PGA	

OPEN SESSION

9:15-9:45	Congress's Charge to Assess Gender Differences Among Science, Engineering, and Math Faculty	**TAB 2**
	Jean Toal Eisen, U.S. Senate, Commerce, Science, and Transportation Committee	
	Rachana Bhowmik, U.S. Senate, Office of Senator Wyden	
	Tamara Jackson, U.S. Senate, Commerce, Science, and Transportation Committee (invited)	
9:45-10:00	Break	
10:00-11:30	Studying Gender Differences Among Science, Engineering, and Math Faculty	**TAB 3**
	John Tsapogas, Alice Hogan, Alan Rapoport, Joan Burrelli, NSF	
	Jerome Bentley, Mathtech	

11:30-12:00	The National Academies Study:	**TAB 4**
	Statement of Task and Project Goals	
	Jong-on Hahm	
12:00-1:00	Lunch	
1:00-2:00	Simultaneous Study by GAO	
	Sonya Harmeyer, Kopp Michelotti, John Mingus,	
	Jim Rebbe, GAO	
2:00-3:00	Surveys of Faculty	**TAB 5**
	Rachel Ivie, AIP	
	Jura Viesulas, ACS	
3:00-3:15	Break	
3:15-4:30	Surveys of Faculty	**TAB 6**
	Sam Rankin III, AMS	
	Tom Price, AAES	
	Susan Hosek, RAND (via conference call)	

CLOSED SESSION

4:30-5:00	Committee Discussion
6:30	Committee Working Dinner

January 30, 2004

OPEN SESSION

8:00-8:30	Continental Breakfast	**TAB 7**
	Selecting dates for next meetings	
8:30-9:30	Building on *From Scarcity to Visibility*	**TAB 8**
	J. Scott Long, Indiana University	
9:30-10:30	Collecting and Using Data from University Studies	**TAB 9**
	Joan Girgus, Princeton University	
	Phoebe Leboy, University of Pennsylvania	
	John Curtis, AAUP	
	Linda Zimbler, NCES	

CLOSED SESSION

10:30-1:00	Committee Discussion of Tasks	**TAB 10**
	Myron Uman, IRB	

Appendix 1-4
The Surveys

The committee designed the surveys to collect information that has gone largely uncollected—or has been done for a few universities, but not across many institutions. As noted earlier in the chapter, the committee designed the departmental survey to focus on processes, particularly tenure, promotion, and hiring, as well as on departmental characteristics. The faculty survey, on the other hand, was designed to assess the resources individual faculty received and to collect sufficient information on faculty to allow for comparisons across fields or by ranks.

The American Institute of Physics (AIP) was contracted to craft the final survey instruments and implement the surveys. The surveys were developed during September 2004. The departmental questionnaire was primarily a mailed instrument. The faculty questionnaire was primarily a Web-based instrument. For both surveys, multiple follow-ups occurred by mail for departments and by e-mail for faculty.

The theoretical population for the departmental chair survey consisted of 534 departments. This represents 89 departments from the 89 Research I institutions multiplied by the six disciplines: biological sciences, chemistry, civil engineering, electrical engineering, mathematics, and physics. In actuality, a few institutions did not offer all six programs. One institution, Rockefeller University, had an organizational structure that seemed very different from the traditional notion of a "department." This school was not included in the survey. As a first step, the committee consulted the institutions' Web sites and identified the names of the six programs. The names of each program and a link to the program's Web site are listed at the conclusion of this summary.[1]

In the case of biology, 87 units were identified. Biology was the most complicated, since it is an evolving discipline. Biology "departments," as thought of in the traditional sense and possessing initial decision-making authority for hiring, tenure and promotions, are called by a variety of names. They are often at least minimally interdisciplinary among the biological sciences, so some units included biochemistry or biophysics; in other cases, the units were subsets of the biological sciences. Departments of molecular and cellular biology are an example of this latter case. In one instance, all the departments had been merged into a single school and so this was included for that institution.

In chemistry, 87 departments were identified. The majority were departments of chemistry, while a few were chemistry and biochemistry. In civil engineering, 69 departments were identified. Often civil engineering was bundled with environmental engineering, and less often with construction engineering, architectural

---

[1] Note that URLs may have changed between the preparation and release of this report.

engineering, or mechanical engineering. In electrical engineering, 77 departments were identified. Electrical engineering departments often included computer engineering. In mathematics, 86 departments were identified. One of the remaining three institutions only offered mathematics as part of an undergraduate college, and so it was excluded. In a few instances, mathematics departments also included statistics. In one case, a joint mathematics and computer science department was included. Finally, in physics, 86 departments were identified. One of the remaining three institutions only offered physics as part of an undergraduate college, and so it was excluded. About half of the departments included astronomy in the department.

The result of this was 492 departments. Initially, the committee's goal was to examine a sample of departments. After further reflection, however, the committee decided a census would be more fruitful. Partly, this reflected a concern that there would be very few responses for women. For example, the questionnaire asked how many faculty were hired in the past 2 years. While many departments were hiring, few hires were women. To increase this latter number, all departments received the sample. Second, the advantage of the census lies in being able to make comparisons between disciplines, e.g., chemistry versus biology, for all Research I institutions.

In all 417 departments responded to the questionnaire. This gives an overall response rate of 85 percent, which is a respectable response. By discipline, electrical engineering had the lowest response rate, while physics had the highest. One might speculate that the fact that AIP sent the survey, and was familiar with physics departments from other survey projects, might have contributed to the higher return for physics departments.

Discipline	Responded	Sample	Percent
Biological sciences	76	87	87
Chemistry	76	87	87
Civil engineering	55	69	80
Electrical engineering	59	77	77
Mathematics	74	86	86
Physics	77	86	90

To generate the faculty sample, the committee collected faculty rosters, for assistant, associate, and full professors, at each of the 492 departments. This was done by consulting each department's Web site for a faculty list. Second, the committee identified the assistant, associate, and full professors in the department. This step was more complex. The committee started with the faculty roster on the individual institution's departmental Web sites. If it identified these three types of faculty, then those faculty members' names were entered into a spreadsheet. If the Web site did not identify faculty members' ranks, then the committee turned to university catalogues. In the event that this failed (because catalogues were not available on line), the committee examined individual faculty members' Web sites.

The following faculty were not included: lecturers, instructors, emeriti professors, research professors, adjunct faculty, visiting faculty, and courtesy appointments. In addition, jointly appointed faculty, where the department in question was the secondary appointment, were not included. Thus, an associate professor of chemistry with a joint appointment in biology, would be counted in chemistry, but not in biology. This process resulted in a final tally of approximately 16,400 faculty.

There are obvious, potential limitations to this approach. Specifically, departmental roster Web sites and college catalogues may be out of date. Recently hired faculty may not have been added to Web sites, while faculty who have left positions might not have been removed. Faculty may have received promotions that have yet to be reflected on departmental Web sites. As a result, it is likely that a few professorial faculty will be missed or misplaced.

Third, the committee identified the gender of each faculty member. This was done primarily by relying on faculty names and photographs on departmental roster Web sites. Where there was some question as to the faculty member's gender, internet research was attempted, and failing that, the department was called. The results of these efforts are captured in the following table.

Population of Faculty in Six Disciplines at Research I Institutions

Department	Gender	Professor	Associate Professor	Assistant Professor	Total
Biology	Male	1222	481	427	2130
	Female	262	176	199	637
Total					2767
Chemistry	Male	1513	331	408	2252
	Female	150	72	101	323
Total					2575
Civil engineering	Male	787	371	302	1460
	Female	57	50	78	185
Total					1645
Electrical engineering	Male	1579	575	531	2685
	Female	79	76	70	225
Total					2910
Mathematics	Male	2153	565	445	3163
	Female	151	76	102	329
Total					3492
Physics	Male	1994	413	407	2814
	Female	119	49	67	235
Total					3049
Total		10066	3235	3137	16438

The committee then took a systematic sample of 50 faculty per gender, rank, and field. Fowler (1993) describes the general procedure: "When drawing a systematic sample from a list, the researcher first determines the number of entries on the list and the number of elements from the list that are to be selected. Dividing the latter by the former will produce a fraction. Thus, if there are 8,500 people on a list and a sample of 100 is required, 1/85 of the list (i.e., 1 out of every 85 persons) is to be included in the sample. In order to select a systematic sample, a start point is designated by choosing a random number from 1 to 85. The randomized start ensures that it is a chance selection process. Given that start, the researcher proceeds to take every 85th person on the list." In some cases, because there are so few women in a particular field at a particular rank, all were selected.[2]

Pre-notice letters were sent to deans/provosts and to department chairs to alert them to the forthcoming questionnaires and also to ask for their assistance and encouragement in filling out the form. Anecdotally, feedback from the administration was positive and encouraging. The departmental census was offered as both a mail-based and Web-based questionnaire. The departmental questionnaire was mailed in November, 2004. A series of follow-ups was undertaken.

The faculty questionnaire was designed as a web-based survey, although some respondents requested a hard copy from the contractor. Faculty received an e-mail request to fill out the survey along with a link to the survey, hosted on the contractor's server.[3] Faculty received multiple e-mail follow-ups.

Some faculty had to be removed or re-classified for various reasons. These included accidental duplication of a faculty member in the sample, faculty member was deceased, information regarding faculty member (i.e., rank) was incorrect, and faculty member was no longer at the institution (and had not moved to another Research I institution). The most frequent problem was that the data on the departmental Web sites was incorrect; usually out of date. The final sample involved 1,834 individuals.

---

[2] The sample was sent to the contractor. Once it was confirmed to have reached the contractor, the original file was deleted. Neither the committee nor the National Academies would know the names of potential respondents to the faculty survey.

[3] Fortunately, almost all e-mails were correct. "Bounce backs," or non-working e-mails, were corrected. It is possible, though, that the wrong e-mail was collected and used, but that the contractor was not aware that this was an incorrect e-mail, and the respondent was never contacted.

Final Sample, Including Respondents, Non-respondents, Refusals, Removals

Department	Gender	Professor	Associate Professor	Assistant Professor	Total
Biology	Male	59	53	42	154
	Female	58	55	44	157
Total					*311*
Chemistry	Male	64	49	43	156
	Female	48	50	44	142
Total					*298*
Civil engineering	Male	61	55	36	152
	Female	44	56	56	156
Total					*308*
Electrical engineering	Male	51	54	51	156
	Female	53	50	45	148
Total					*304*
Mathematics	Male	69	43	43	155
	Female	53	46	44	143
Total					*298*
Physics	Male	61	42	50	153
	Female	58	48	56	162
Total					*315*
Total		*679*	*601*	*554*	*1834*

Of these 1,834 individuals, 91 had to be removed from the sample, because they should not have been included in the population (e.g., were deceased, no longer at a Research I institution, or not one of the three professorial ranks). Overall, 41 men and 50 women or 24 professors, 29 associate professors, and 38 assistant professors were removed.

Individuals Removed from Sample

Department	Gender	Professor	Associate Professor	Assistant Professor	Total
Biology	Male	2	5	1	8
	Female	1	3	8	12
Total					20
Chemistry	Male	1	2	1	4
	Female	1	1	4	6
Total					10
Civil engineering	Male	1	3	2	6
	Female	1	1	0	2
Total					8
Electrical engineering	Male	0	0	0	0
	Female	2	3	2	7
Total					7
Mathematics	Male	5	4	8	17
	Female	4	2	7	13
Total					30
Physics	Male	0	2	4	6
	Female	6	3	1	10
Total					16
Total		24	29	38	91

Approximately, 1,743 individuals made up the corrected sample. Of these 1,347 responded to the questionnaire. Additionally, 1,278 filled out the survey, while 69 individuals responded by refusing to complete the survey.

Respondents (Including Those Who Responded by Refusing)

Department	Gender	Professor	Associate Professor	Assistant Professor	Total
Biology	Male	46	33	34	113
	Female	49	44	31	124
Total					237
Chemistry	Male	51	34	32	117
	Female	39	41	32	112
Total					229
Civil engineering	Male	40	38	26	104
	Female	31	48	48	127
Total					231
Electrical engineering	Male	35	31	42	108
	Female	40	39	31	110
Total					218
Mathematics	Male	44	25	25	94
	Female	35	36	27	98
Total					192
Physics	Male	50	34	30	114
	Female	41	34	51	126
Total					240
Total		501	437	409	1347

Non-respondents

Department	Gender	Professor	Associate Professor	Assistant Professor	Total
Biology	Male	11	15	7	33
	Female	8	8	5	21
Total					*54*
Chemistry	Male	12	13	10	35
	Female	8	8	8	24
Total					*59*
Civil engineering	Male	20	14	8	42
	Female	12	7	8	27
Total					*69*
Electrical engineering	Male	16	23	9	48
	Female	11	8	12	31
Total					*79*
Mathematics	Male	20	14	10	44
	Female	14	8	10	32
Total					*76*
Physics	Male	11	6	16	33
	Female	11	11	4	26
Total					*59*
Total		154	135	107	*396*

To conclude:

- 1,834 individuals comprised the sample.
- 1,743 individuals comprised the corrected sample (excludes removals).
- 1,347 individuals responded (includes refusals).
- 1,278 individuals provided some data.
- 396 individuals did not respond.

The response rate for the survey (number of completed questionnaires divided by number of valid sample elements) is 1,278/1,743 or 73 percent.

Immediately following this text are the list of 492 departments surveyed, the departmental questionnaire, and the faculty questionnaire.

Appendix 1-5
Survey Instruments

THE NATIONAL ACADEMIES
2004-05 Faculty Hiring, Tenure & Promotion Questionnaire

All responses will be kept confidential. Contact information will be used for follow-up purposes only.

Survey completed by:

Name _____

Title _____

Phone _____

E-mail _____

Section A: Department Characteristics

A1. Over the past ten years, how many chairs or heads has your department had?

_____ Males

_____ Females

A2. How many endowed chairs or named professorships does your department have?

_____ Held by males

_____ Held by females

A3. Does your department or university have the following? Check all that apply.

☐ Child care center

☐ Policy that permits hiring spouses/partners

☐ Mortgage/housing assistance

☐ Paid family leave/workload relief

☐ Family leave longer than the 12 weeks required by law

☐ Job placement assistance for spouses/partners

☐ Other family policies

_____

_____

A4. Are terms at your institution on the semester or quarter system?

☐ Semester

☐ Quarter

☐ Other

A5. How many faculty are at each of the following rank in your department?

	Males	Females
Tenured Full Professors....................	_____	_____
Tenured Associate Professors..................	_____	_____
Tenure-Track Associate Professors..................	_____	_____
Tenured Assistant Professors..................	_____	_____
Tenure-Track Assistant Professors..................	_____	_____
Full-time Faculty, not on Tenure Track............	_____	_____
Total............................	_____	_____

Continue to page 2.

Section B: Tenure Process

B1. During the two previous academic years (2002-03 and 2003-04), how many tenure-track faculty left your department without being granted tenure?

———— Males

———— Females

B2. During the two previous academic years (2002-03 and 2003-04), how many tenured faculty left your department to go to another institution?

———— Males

———— Females

B3. How many years after entering a tenure-track position do untenured faculty typically come up for tenure in your department?

———— Years

B4. Are tenure and promotion to associate professor granted in a single decision?

☐ No

☐ Yes

B5. Does your department or university have a formal family or personal leave policy that allows stopping or extending the tenure clock?

☐ No

☐ Yes

    How many times can the clock be stopped or extended? ————

    For how many years can the clock be stopped or extended each time? ————

B6. At your university, which steps are part of the formal procedure for granting tenure? Check all that apply.

☐ Letters are solicited by the department.

☐ Letters are solicited by a committee outside the department.

☐ Department makes a recommendation.

☐ Dean reviews and makes recommendation.

☐ School or college committee reviews and makes recommendation.

☐ University committee reviews and makes recommendation.

☐ President or Provost reviews and makes recommendation

☐ Other step(s) ————————

B7. How does an untenured professor learn about the steps in the tenure process? Check all that apply.

☐ Written procedures from the university

☐ Written procedures from the department

☐ Information communicated orally by chair or staff

☐ Informal networks

☐ Other

————————————

B8. During the two previous academic years (2002-03 and 2003-04), were any faculty in your department considered for tenure?

☐ No

☐ Yes

    How many in 2002-03? ————

    How many in 2003-04? ————

If you considered at least one faculty member for tenure during 2002-03 and 2003-04, continue to page 3.

If you did not consider any faculty member for tenure during 2002-03 and 2003-04, skip to page 4.

2

Section B: Tenure Process

Please answer the following questions for each faculty member who was considered for tenure during 2002-03 and 2003-0
If you considered more than four faculty members, please copy this page and attach the additional sheet(s).

Year of Decision

☐ 2002-03   ☐ 2003-04

The faculty member being considered was

☐ Male   ☐ Female

Tenure was

☐ Granted   ☐ Not granted

Number of Tenure Committee members

_____ Males   _____ Females

The committee chair was

☐ Male   ☐ Female

Year of Decision

☐ 2002-03   ☐ 2003-04

The faculty member being considered was

☐ Male   ☐ Female

Tenure was

☐ Granted   ☐ Not granted

Number of Tenure Committee members

_____ Males   _____ Females

The committee chair was

☐ Male   ☐ Female

Year of Decision

☐ 2002-03   ☐ 2003-04

The faculty member being considered was

☐ Male   ☐ Female

Tenure was

☐ Granted   ☐ Not granted

Number of Tenure Committee members

_____ Males   _____ Females

The committee chair was

☐ Male   ☐ Female

Year of Decision

☐ 2002-03   ☐ 2003-04

The faculty member being considered was

☐ Male   ☐ Female

Tenure was

☐ Granted   ☐ Not granted

Number of Tenure Committee members

_____ Males   _____ Females

The committee chair was

☐ Male   ☐ Female

3

Continue to page 4.

Section C: Promotions to Full Professor

C1. At your university, which steps are part of the formal procedure for promoting faculty to full professor? Check all that apply.

- ☐ Letters are solicited by the department.
- ☐ Letters are solicited by a committee outside the department.
- ☐ Department makes a recommendation.
- ☐ Dean reviews and makes recommendation.
- ☐ School or college committee reviews and makes recommendation.
- ☐ University committee reviews and makes recommendation.
- ☐ President or Provost reviews and makes recommendation.
- ☐ Other step(s)

  _____

  _____

C2. During the two previous academic years (2002-03 and 2003-04), were any faculty in your department considered for promotion to full professor?

- ☐ No
- ☐ Yes

  ↳ How many in 2002-03? _____

  How many in 2003-04? _____

If you considered at least one faculty member for promotion to full professor during 2002-03 and 2003-04, please complete the following questions for each faculty member considered. If you considered more than two, please copy this page and attach the additional sheet(s).

If you did not consider any faculty member for promotion to full professor during 2002-03 and 2003-04, skip to page 5.

Year of Decision

  ☐ 2002-03   ☐ 2003-04

The faculty member being considered was

  ☐ Male   ☐ Female

Promotion to full professor was

  ☐ Granted   ☐ Not granted

Number of Promotion Committee members

  _____ Males   _____ Females

The committee chair was

  ☐ Male   ☐ Female

Year of Decision

  ☐ 2002-03   ☐ 2003-04

The faculty member being considered was

  ☐ Male   ☐ Female

Promotion to full professor was

  ☐ Granted   ☐ Not granted

Number of Promotion Committee members

  _____ Males   _____ Females

The committee chair was

  ☐ Male   ☐ Female

4

Continue to page 5.

Section D: Hiring and Recruitment

D1. In addition to your department's procedures, which steps _outside your department_ are part of the formal procedure for hiring tenured or tenure-track faculty? Check all that apply.

☐ Search committee includes members from outside the department.

☐ Before interviewing, there is a review of candidates by the Dean or by a committee external to the dept.

☐ Institutional funds are made available for the search.

☐ Dean approves hiring recommendation.

☐ Other step(s)

_____

D2. What steps (if any) has your department or institution taken to increase the gender diversity of your candidate pool?

_____
_____
_____
_____
_____

D3. During the two previous academic years (2002-03 and 2003-04), did your department recruit any new tenured or tenure-track faculty?

☐ No

☐ Yes

　　↳ How many in 2002-03?　_____

　　　How many in 2003-04?　_____

If you recruited for at least one tenured or tenure-track position during 2002-03 and 2003-04, please complete the following questions for each recruitment.

Space for additional recruitments is on pages 6 and 7. If you recruited for more than four total positions, please copy page 6 and attach the additional sheet(s).

If you did not recruit any faculty during 2002-03 and 2003-04, skip to page 7.

Year of recruitment　　☐ 2002-03　　☐ 2003-04

Search was initiated as

　　☐ Tenured

　　☐ Tenure-track

Number of applicants or candidates for the position (estimate)

　　Total............　_____

　　Males ....　_____

　　Females..........　_____

　　Don't Know....　_____

Number of applicants interviewed on campus

　　Total............　_____

　　Males ....　_____

　　Females..........　_____

First offer was made to...　☐ A male　　☐ A female

Was someone hired?

　　☐ Yes, into a tenured position

　　☐ Yes, into a tenure-track position

　　☐ Yes, into a temporary position

　　☐ No

Who was hired?　　☐ A male　　☐ A female

New faculty member's research is...

　　☐ Theoretical

　　☐ Experimental or applied

　　☐ Other

_____

New faculty member's start-up package was (exclusive of renovations)...

　　$ _____

Number of Search Committee members

　　_____ Males　　_____ Females

The committee chair was

　　☐ Male　　☐ Female

5

| Year of recruitment | ☐ 2002-03 | ☐ 2003-04 | Year of recruitment | ☐ 2002-03 | ☐ 2003-04 |

Search was initiated as

   ☐ Tenured

   ☐ Tenure-track

Number of applicants or candidates for the position (estimate)

   Total...............   ———

   Males ....   ———

   Females... .....   ———

   Don't Know....   ———

Number of applicants interviewed on campus

   Total...............   ———

   Males ....   ———

   Females........   ———

First offer was made to... ☐ A male   ☐ A female

Was someone hired?

   ☐ Yes, into a tenured position

   ☐ Yes, into a tenure-track position

   ☐ Yes, into a temporary position

   ☐ No

Who was hired? ☐ A male   ☐ A female

New faculty member's research is...

   ☐ Theoretical

   ☐ Experimental or applied

   ☐ Other

   _____

New faculty member's start-up package was (exclusive of renovations)...

   $ _____

Number of Search Committee members

   ——— Males   ——— Females

The committee chair was

   ☐ Male   ☐ Female

Search was initiated as

   ☐ Tenured

   ☐ Tenure-track

Number of applicants or candidates for the position (estimate)

   Total...............   ———

   Males ....   ———

   Females........   ———

   Don't Know....   ———

Number of applicants interviewed on campus

   Total...............   ———

   Males ....   ———

   Females........   ———

First offer was made to... ☐ A male   ☐ A female

Was someone hired?

   ☐ Yes, into a tenured position

   ☐ Yes, into a tenure-track position

   ☐ Yes, into a temporary position

   ☐ No

Who was hired? ☐ A male   ☐ A female

New faculty member's research is...

   ☐ Theoretical

   ☐ Experimental or applied

   ☐ Other

   _____

New faculty member's start-up package was (exclusive of renovations)...

   $ _____

Number of Search Committee members

   ——— Males   ——— Females

The committee chair was

   ☐ Male   ☐ Female

Continue to page 7.

Year of recruitment ☐ 2002-03 ☐ 2003-04

Search was initiated as

☐ Tenured

☐ Tenure-track

Number of applicants or candidates for the position (estimate)

Total............ _____

Males .... _____

Females........ _____

Don't Know.... _____

Number of applicants interviewed on campus

Total............. _____

Males .... _____

Females......... _____

First offer was made to... ☐ A male ☐ A female

Was someone hired?

☐ Yes, into a tenured position

☐ Yes, into a tenure-track position

☐ Yes, into a temporary position

☐ No

Who was hired? ☐ A male ☐ A female

New faculty member's research is...

☐ Theoretical

☐ Experimental or applied

☐ Other

New faculty member's start-up package was (exclusive of renovations)...

$ _____

Number of Search Committee members

_____ Males _____ Females

The committee chair was

☐ Male ☐ Female

Section E: Final Remarks

E1. What barriers do you experience in recruitment and retention of tenured and tenure-track faculty?

_____
_____
_____
_____
_____
_____

E2. What barriers do you experience in recruitment and retention of _female_ tenured and tenure-track faculty?

_____
_____
_____
_____
_____
_____

E3. Do you have any additional comments you wish to share with the National Research Council?

_____
_____
_____
_____
_____
_____

7

Continue to page 8.

Participation in this survey is voluntary. If you participate, your response will be kept confidential:

- *Contact information will be used only for follow-up purposes. No information from your response that could be used to discern your identity or how you responded to any survey question will be made available to anyone not involved in the collection and compilation of survey results.*

- *When data collection and compilation has been completed, all information in your response that could be used to discern your identity or how you responded to any survey question will be destroyed.*

- *Your response will be compiled with the responses from other survey participants. Only aggregated information that cannot be used to discern the identity of any survey participant or how they responded to any survey question will be used in any report or presentation concerning the survey or made available to the public.*

THE NATIONAL ACADEMIES
2004-05 Faculty Questionnaire

- *Participation is voluntary and all responses will be kept confidential.*
- *Your name will NOT be associated with your answers, and identifying information will be removed.*
- *Full explanation of confidentiality.*

A1. What is your current rank at this institution?

- ○ Full Professor
- ○ Associate Professor
- ○ Assistant Professor
- ○ Research Professor
- ○ Other [   ]

A2. As of the Fall 2004 term, were you employed full-time or part-time at this institution?

- ○ Full-time
- ○ Part-time

A3. What date were you hired by this institution?

| Choose One | Month | [   ] Year |

A4. What is your highest degree?

- ○ PhD
- ○ Other [   ]

A5. When did you receive your highest degree?

| Choose One | Month | [   ] Year |

A6. <u>For engineering faculty only</u>: Do you hold a P.E.?

○ No

○ Yes

A7. What is your current tenure status?

○ Tenured

○ Tenure-track

○ Not on tenure track

[continue]

B1. Were you hired at this institution with tenure?

○ Yes

○ No

If no, when were you granted tenure at this institution?

| Choose One | Month | | Year |

B2. Have you received an offer to leave your current institution in the last five years?

○ No

○ Yes, one offer

○ Yes, more than one offer

If yes, why did you stay at your current institution? Check all that apply.

☐ I am happy at my current institution.

☐ The other offer was not as good.

☐ My current institution made a counteroffer.

☐ I prefer my current geographic location.

☐ I didn't want to relocate due to family reasons.

☐ Other

continue

C1. Do you have a faculty mentor at this institution?

  ● *If you have tenure, please answer about the time before you had tenure.*

○ No

○ Yes

  If yes, how often do/did you interact with this mentor?

  ○ Very frequently

  ○ Somewhat frequently

  ○ Somewhat infrequently

  ○ Very infrequently

  ○ Never

C2. Were you ever a postdoctoral fellow?

○ No

○ Yes

  If yes, for how many years?

  ▢▢▢▢ Years

C3. How were you made aware of your institution's policy on tenure?

	Yes	No
I was given a written policy.	○	○
The chair or administration told me about the policy.	○	○
I learned about the policy from other faculty.	○	○
Other...	○	○

▭

C4. Have you ever stopped or extended your tenure clock at this institution?

○ No

○ Yes

If yes, for what reason(s)?

	Yes	No
Childbirth or adoption	○	○
Other family issues	○	○
Personal health	○	○
Other	○	○

continue

D1. Have you been promoted in rank at this institution?

○ No

○ Yes

If yes, what was your rank when you were hired?

○ Associate Professor

○ Assistant Professor

○ Research Faculty or Staff

○ Visiting Professor

○ Other [_____]

Please tell us about your promotion(s) at this institution.

Date of promotion	Date of promotion
Choose One Month [____] Year	Choose One Month [____] Year
New rank:	New rank:
[_____]	[_____]

Date of promotion	Date of promotion
Choose One Month [____] Year	Choose One Month [____] Year
New rank:	New rank:
[_____]	[_____]

D2. Is your research primarily...

Mathematics Faculty Only	**Faculty in All Other Disciplines**
○ Pure	○ Theoretical
○ Applied	○ Experimental
○ Other	○ Other
[_____]	[_____]

How many articles in refereed <u>journals</u> have you published since November 2002?

- *Include only articles that have been accepted for publication.*

[   ] Number sole-authored
[   ] Number co-authored

How many papers in refereed <u>conference proceedings</u> have you had accepted since November 2002?

- *Include only papers that were refereed.*

[   ] Number sole-authored
[   ] Number co-authored

D4. Have you been nominated by your current department or institution for any international or national prizes or awards?

○ No
○ Yes

If yes, please give the dates of your three most recent nominations.

[   ] Year

[   ] Year

[   ] Year

D5. Approximately how much lab space do you have at this institution?

[   ] Square feet   ☐ N/A

D6. Do you work with others at this institution on a research team?

○ No

○ Yes

   If yes, approximately how many people work on your primary research team?

   • *Include only those at your institution.*

   [＿＿＿] Tenured or tenure-track faculty more senior than you

   [＿＿＿] Tenured or tenure-track faculty more junior than you

   [＿＿＿] Research faculty or staff

   [＿＿＿] Postdocs

   [＿＿＿] Graduate students

   [＿＿＿] Total

D7. Do you have access to equipment that you need to conduct your research?

○ Yes, I have everything I need.

○ I have most of what I need.

○ I do not have access to major pieces of equipment that I need for my research.

○ N/A

D8. Do you have access to clerical support that you need for your teaching and research?

○ Yes, I have access to all the clerical support I need.

○ Yes, I have access to some clerical support, but could use more.

○ No, I do not have access to clerical support.

○ N/A

D9. How many of the following do you directly supervise this semester or term?

[＿＿＿] Graduate research assistants

[＿＿＿] Graduate teaching assistants

[＿＿＿] Postdocs   ☐   N/A (do not have postdocs in this field)

D10. What is the total dollar amount of the research grants on which you served as Principal Investigator or Co-Principal Investigator during the 2004-05 academic year?

- *Include only direct costs for academic year 2004-05.*

$ [   ] .00

D11. During this semester or term, how many individuals at this institution other than yourself are supported by all the grants on which you were PI or Co-PI?

- *Include students, postdocs, faculty and support staff*

[   ] Number of individuals (do not include yourself)

D12. During this semester or term, how many classes are you teaching?

[   ] Undergraduate classes
[   ] Graduate classes

If you are teaching classes, how many different preparations do these classes represent?

[   ] Number of preparations

If you are teaching classes, approximately how many students do you have in your classes this semester or term?

[   ] Undergraduates

[   ] Graduates

D13. During the last five years, has your teaching load been reduced because of grant funding that you received?

○ No
○ Yes

D14. What is your base salary at this institution?

$ [   ] .00 Over... ○ 9-10 months
                          ○ 11-12 months

D15. During the last five years, have you been given travel money by your department or institution to attend professional conferences or to conduct research off site?

- *Do not include travel money that was part of a grant on which you were PI or Co-PI.*

○ No
○ Yes

D16. Over the past five years, please indicate your role on the following committees.

	Did not participate	Served on	Chaired
Departmental committees			
Undergraduate curriculum	○	○	○
Graduate curriculum	○	○	○
Executive	○	○	○
Promotion and tenure	○	○	○
Faculty search	○	○	○
Fellowship	○	○	○
Graduate admissions	○	○	○
Facilities or space	○	○	○
Program review	○	○	○
Other	○	○	○

	Did not participate	Served on	Chaired
School, college, or university level committees			
Undergraduate curriculum	○	○	○
Graduate curriculum	○	○	○
Department chair or unit head search	○	○	○
Other	○	○	○

D17. During the current semester or term, on how many <u>undergraduate</u> thesis or honors committees did you serve at this institution?

[ ] I served on

[ ] I chaired

D18. During the current semester or term, on how many <u>graduate</u> thesis or honors committees did you serve at this institution?

[ ] I served on

[ ] I chaired

D19. For the current semester or term, what proportion of your time did you spend on each activity? Write a percentage on each line. Give your best estimate. If none, write in "0".

[ ] % Teaching and advising undergraduate students

[ ] % Teaching and advising graduate students

[ ] % Research or scholarship

[ ] % Seeking funding

[ ] % Administration or committee work

[ ] % Service outside the university

[ ] % Outside consulting

[ ] % Other: [   ]

100% Total

D20. How aware are you of your institution's policy on each of the following?

	I know the policy	There is a policy, but I don't know it	I don't know if there is a policy	There is no policy
Tenure	○	○	○	○
Promotion	○	○	○	○
Service requirements	○	○	○	○
Changing the tenure clock	○	○	○	○
Family leave	○	○	○	○
Child care	○	○	○	○
Spousal employment	○	○	○	○
Workload reduction for family reasons	○	○	○	○

D21. Over the past year, with how many faculty members at your institution did you discuss each of the following items?

	Zero	One	Two	Three or More	N/A
Teaching	○	○	○	○	○
Research	○	○	○	○	○
Funding	○	○	○	○	○
Interaction with other faculty members	○	○	○	○	○
Interaction with administration	○	○	○	○	○
Climate in the department	○	○	○	○	○
Personal life	○	○	○	○	○
Family obligations	○	○	○	○	○
Salary	○	○	○	○	○
Benefits	○	○	○	○	○

D22. Are you considering leaving or retiring from your current institution?

○ No, I am not considering leaving or retiring

○ Yes, during this academic year

○ Yes, during the next academic year (2005-06)

○ Yes, some time after the next academic year

If yes, which of the following would be factors in your decision?

	Yes	No
Pay	○	○
Benefits	○	○
Promotion opportunities	○	○
Concerns about tenure	○	○
Planning to retire	○	○
Funding opportunities	○	○
Family-related reasons	○	○
Job location	○	○
Collegiality of faculty	○	○
	Yes	No
Reputation of department or university	○	○
Quality of research facilities	○	○
Access to research facilities	○	○
Opportunities for research collaboration	○	○
Desire to build or lead a new program or area of research	○	○
Teaching and advising	○	○
Pressure to publish	○	○
Start-up package	○	○
Personal health	○	○
Other:	○	○

continue

E1. When you were first hired at this institution, how much were you given in start-up funds?

$ [____] .00   For equipment   ☐ I don't know

$ [____] .00   For renovation of lab space   ☐ I don't know

$ [____] .00   For staff, including postdocs and research associates   ☐ I don't know

$ [____] .00   For other   ☐ I don't know

$ [____] .00   Total

E2. When you were first hired at this institution, were you given lab space?

○ No

○ Yes, I was given the same amount of space I have now.

○ Yes, but I have more lab space now than I did then.

○ Yes, but I have less lab space now than I did then.

○ N/A

E3. How many graduate students, postdocs, and research faculty or staff who were funded by the department or university did your start-up package include?

- *Do not include graduate students, postdocs, or research faculty or staff that you funded on a grant on which you were the PI or Co-PI.*

[____] Graduate research assistants   ☐ or none

[____] Postdocs   ☐ or none

[____] Research faculty or staff   ☐ or none

E4. When you were first hired at this institution, were you given access to equipment that you needed to conduct your research?

- *Do not include equipment that you funded from grants on which you were the PI or Co-PI.*

○ Yes, I was given access to all the equipment I needed.

○ Yes, I was given access to some of the equipment I needed.

○ No, I was not given access to equipment that I needed.

○ N/A

E5. When you were first hired at this institution, were you given the following items?

- *Do not include items that you funded from grants on which you were the PI or Co-PI.*

	Yes	No
Travel funds	○	○
Summer salary	○	○
Moving expenses	○	○

E6. When you were first hired at this institution, were you given a reduced teaching load?

○ No

○ Yes

    If yes, how long was your teaching load reduced?

      [＿＿＿] Number of... ○ Semesters

                         ○ Quarters

E7. What were your main considerations in deciding to work for your current institution?

	Yes	No
Pay	○	○
Benefits	○	○
Promotion opportunities	○	○
Start-up package	○	○
Funding opportunities	○	○
Family-related reasons	○	○
Job location	○	○
Collegiality of faculty	○	○
Reputation of department or university	○	○
Quality of research facilities	○	○
Access to research facilities	○	○
Opportunities for research collaboration	○	○
Desire to build or lead a new program or area of research	○	○
This was the only offer I received.	○	○
Other	○	○

continue

F1. Are you...

- ○ Male
- ○ Female

F2. What was your marital status in the Fall 2004 semester or term?

- ○ Single, divorced, widowed, or separated
- ○ Married or in a marriage-like relationship

- *If married or in a marriage-like relationship, please answer the following questions.*

During the Fall 2004 term, was your spouse or significant other employed in a professional position at a college, university, or affiliated research institute?

- ○ No
- ○ Yes

Did your institution assist in finding your spouse or significant other employment?

- ○ No
- ○ Yes

During the Fall 2004 term, did you and your spouse or significant other maintain separate households because of employment?

- ○ No
- ○ Yes

F3. If you have children, please tell us how many you have..

- [ ] Children under age 7
- [ ] Children age 7-12
- [ ] Children age 13-17
- [ ] Children age 18 or older

- ☐ I do not have children

F4. Do you have dependent children who live with you most of the time?

○ No

○ Yes

F5. Are you primarily responsible for the care of relatives other than children?

○ No

○ Yes

F6. Does your institution offer on-site child care?

○ No

○ Yes

○ Don't know

F7. Since receiving your highest degree, how many years did you spend employed outside academia?

[ ] Years

F8. Since receiving your highest degree, how long did you spend not employed?

[ ] Months   [ ] Years

F9. Do you have any additional comments you would like to share with the National Academies?

submit

Press submit to finish the questionnaire.

Appendix 1-6
Departments in Survey

BIOLOGY

Name of Institution	Name of Department	URL
Arizona State University	School of Life Sciences	http://sols.asu.edu/people/faculty.php
Boston University	Biology	http://www.bu.edu/biology/faculty_and_staff.html
Brown University	Molecular, Cell and Biochemistry	http://www.brown.edu/Departments/Molecular_Biology/faculty.html
California Institute of Technology	Biology	http://biology.caltech.edu/faculty/
Carnegie-Mellon University	Biological Sciences	http://www.cmu.edu/bio/contacts/faculty.shtml
Case Western Reserve University	Biology	http://www.case.edu/artsci/biol/
Colorado State University	Biology	http://rydberg.biology.colostate.edu/faculty/
Columbia University	Biological Sciences	http://www.columbia.edu/cu/biology/pages/fac/main/intro/index.html
Cornell University	Molecular Biology and Genetics	http://www.mbg.cornell.edu/cals/mbg/faculty-staff/faculty/index.cfm
Duke University	Biology	http://fds.duke.edu/db/aas/Biology/faculty/photos.html
Emory University	Biology	http://www.biology.emory.edu/
Florida State University	Biological Science	http://www.bio.fsu.edu/faculty.php
Georgetown University	Biology	http://biology.georgetown.edu/
Georgia Institute of Technology	Biology	http://www.biology.gatech.edu/faculty/
Harvard University	Molecular and Cellular	http://www.mcb.harvard.edu/Faculty/List.asp
Howard University	Biology	http://www.biology.howard.edu/Faculty/Faculty.html

BIOLOGY

Name of Institution	Name of Department	URL
Indiana University at Bloomington	Biology	http://www.bio.indiana.edu/facultyresearch/index.html
Iowa State University	Biochemistry, Biophysics, and Molecular	http://www.bb.iastate.edu/FacultyResearchFolder/FacultyListFrameset/FacultyListFrameset.html
Johns Hopkins University	Biology	http://www.bio.jhu.edu/
Louisiana State University	Biological Sciences	http://www.biology.lsu.edu/faculty_listings/abc_fac.html
Massachusetts Institute of Technology	Biology	http://web.mit.edu/biology/www/facultyareas/viewalpha.html
Michigan State University	Biochemistry and Molecular Biology	http://www.bch.msu.edu/faculty/faculty.html
New Mexico State University	Biology	http://biology-web.nmsu.edu/Faculty&Staff/Faculty.htm
New York University	Biology	http://www.nyu.edu/fas/dept/biology/faculty/index.html
North Carolina State University	Biological Sciences	http://harvest.cals.ncsu.edu/biology/index.cfm?pageID=1523
Northwestern University	Biochemistry, Molecular, and Cell	http://www.biochem.northwestern.edu/faculty_and_staff/faculty/
Ohio State University	Molecular Genetics	http://www.osumolgen.org/faculty/
Oregon State University	Microbiology	http://microbiology.science.oregonstate.edu/
Pennsylvania State University	Biology	http://www.bio.psu.edu/home
Princeton University	Molecular Biology	http://www.molbio.princeton.edu/index.php?option=com_content&task=view&id=196&Itemid=244
Purdue University	Biological Science	http://www.bio.purdue.edu/people/faculty/index.php
Rockefeller University	NA	NA

BIOLOGY

Name of Institution	Name of Department	URL
Rutgers, the State University of New Jersey	Molecular Biology and Biochemistry	http://mbb.rutgers.edu/dept/faculty.html
Stanford University	Biological Science	http://www.stanford.edu/dept/biology/faculty.html
State University of New York at Buffalo	Biological Science	http://wings.buffalo.edu/academic/department/fnsm/biosci/staffdir.html
State University of New York at Stony Brook	Biochemistry and Cell Biology	http://www.sunysb.edu/biochem/BIOCHEM/faclist.html
Temple University	Biology	http://www.temple.edu/biology/
Texas A&M University	Biology	http://www.bio.tamu.edu/phone/facphone.htm
Tufts University	Biology	http://ase.tufts.edu/biology/default.asp
Tulane University	Cellular and Molecular Biology	http://tulane.edu/sse/cmb/index.cfm
University of Alabama at Birmingham	Biology	http://www.uab.edu/uabbio/by2.htm
University of Arizona	Molecular and Cellular	http://www.mcb.arizona.edu/directory.cfm?code=P
University of California-Berkeley	Molecular and Cellular	http://mcb.berkeley.edu/cgi-bin/directory.cgi?subroutine=list&listtype=faculty
University of California-Davis	Section of Molecular and Cellular Biology	http://biosci2.ucdavis.edu/BioSci/FacultyAndResearch/DisplayAlphaListingOfFaculty.cfm
University of California-Irvine	Department of Molecular Biology and Biochemistry	http://www.bio.uci.edu/faculty/index.cfm
University of California-Los Angeles	Molecular, Cell and Development Biology	http://www.mcdb.ucla.edu/
University of California-San Diego	Biological Science	http://www.biology.ucsd.edu/bioresearch/list_alpha.html

BIOLOGY

Name of Institution	Name of Department	URL
University of California-San Francisco	Biochemistry and Biophysics	http://biochemistry.ucsf.edu/
University of California-Santa Barbara	Molecular, Cellular, and Developmental Biology	http://www.lifesci.ucsb.edu/mcdb/faculty/faculty.php
University of Chicago	Molecular Genetics and Cell Biology	http://mgcb.bsd.uchicago.edu/
University of Cincinnati	Biological Science	http://www.biology.uc.edu/general.asp?subject=faculty
University of Colorado	Molecular, Cellular, and Developmental Biology	http://mcdb.colorado.edu/
University of Connecticut	Molecular and Cellular Biology	http://mcb.uconn.edu/faculty.php
University of Florida	Molecular Cell Biology	http://www.med.ufl.edu/anatomy/mcb/index.cgi
University of Georgia	Biochemistry and Molecular Biology	http://www.bmb.uga.edu/home/people/index.htm
University of Hawaii at Manoa	Biology	http://www.hawaii.edu/biology/faculty.htm
University of Illinois at Chicago	Biological Science	http://www.uic.edu/depts/bios/
University of Illinois at Urbana-Champaign	Cellular and Molecular Biology	http://www.life.uiuc.edu/mcb/research/ABC_list.html
University of Iowa	Biological Science	http://www.biology.uiowa.edu/directory.php
University of Kansas	Molecular Bioscience	http://www.molecularbiosciences.ku.edu/faculty/index.html
University of Kentucky	Biology	http://biology.uky.edu/sbs/faculty.htm
University of Maryland at College Park	Biology	http://www.life.umd.edu/biology/facultyalpha.html
University of Massachusetts at Amherst	Biology	http://www.bio.umass.edu/biology/faculty/index.phtml
University of Miami	Biology	http://www.bio.miami.edu/people.html

BIOLOGY

Name of Institution	Name of Department	URL
University of Michigan	Molecular, Cellular, and Developmental Biology	http://www.mcdb.lsa.umich.edu/index.php
University of Minnesota, Twin Cities	Biochemistry, Molecular and Biophysics	http://biosci.cbs.umn.edu/BMBB/faculty/index.shtml
University of Missouri-Columbia	Biological Science	http://www.biology.missouri.edu/
University of Nebraska-Lincoln	Biological Science	http://www.biosci.unl.edu/faculty/
University of New Mexico	Biology	http://biology.unm.edu/
University of North Carolina at Chapel Hill	Biology	http://www.bio.unc.edu/faculty/
University of Pennsylvania	Biology	http://www.bio.upenn.edu/faculty/
University of Pittsburgh	Biological Science	http://www.pitt.edu/~biology/
University of Rochester	Biology	http://www.rochester.edu/College/BIO/index.php
University of Southern California	Molecular and Computational Biology	http://college.usc.edu/bisc/molecular/home/
University of Tennessee, Knoxville	Biology	http://web.bio.utk.edu/bcmb/Faculty.htm
University of Texas at Austin	Molecular, Cell and Developmental Biology	http://www.biosci.utexas.edu/mcdb/
University of Utah	Genetics and Molecular Biology	http://www.biology.utah.edu/faculty.php
University of Virginia	Biology	http://www.virginia.edu/biology/New_Bio_Home_files/directory/faculty_directory.html
University of Washington	Biology	http://protist.biology.washington.edu/bio2/people/faculty.html
University of Wisconsin-Madison	Biochemistry	http://www.biochem.wisc.edu/faculty/index.html
Utah State University	Biology	http://www.biology.usu.edu/
Vanderbilt University	Biological Science	http://sitemason.vanderbilt.edu/biosci/index

BIOLOGY

Name of Institution	Name of Department	URL
Virginia Commonwealth University	Biology	http://www.has.vcu.edu/bio/people/index.html#faculty
Virginia Polytechnic Institute & State University	Biology	http://www.biol.vt.edu/faculty/index.html
Washington University	Biology	http://www.biology.wustl.edu/faculty/index.php
Wayne State University	Biological Science	http://www.clas.wayne.edu/unit-faculty.asp?UnitID=4
West Virginia University	Biology	http://www.as.wvu.edu/biology/faculty/faculty.htm
Yale University	Cellular Molecular and Developmental Biology	http://www.yale.edu/mcdb/facultystaff/index.html
Yeshiva University	Developmental and Molecular Biology	http://www.yu.edu/aecomdb/facultydir/academictest2.asp?id=001#start

CHEMISTRY

Name of Institution	Name of Department	URL
Arizona State University	Chemistry and Biochemistry	http://chemistry.asu.edu/graduate/facultyResearch.asp
Boston University	Chemistry	http://www.bu.edu/chemistry/faculty/
Brown University	Chemistry	http://www.chem.brown.edu/people/
California Institute of Technology	Chemistry and Chemical Engineering	http://www.cce.caltech.edu/faculty/index.html
Carnegie-Mellon University	Chemistry	http://www.chem.cmu.edu/faculty/fac-profiles.html
Case Western Reserve University	Chemistry	http://www.case.edu/artsci/chem/faculty/

CHEMISTRY

Name of Institution	Name of Department	URL
Colorado State University	Chemistry	http://129.82.76.41:591/chemdept2002/FMPro?-db=faculty-staff2002.fp5&-format=fac_alpha-fordatabasesite.html&-lay=Layout1&-op=eq&category=faculty&-sortfield=lastname&-sortorder=ascending&-Max=All&-find
Columbia University	Chemistry	http://www.columbia.edu/cu/chemistry/fac-rch/faculty/index.html
Cornell University	Chemistry and Chemical Biology	http://www.chem.cornell.edu/faculty/index.asp
Duke University	Chemistry	http://fds.duke.edu/db/aas/Chemistry/faculty
Emory University	Chemistry	http://www.emory.edu/CHEMISTRY/faculty/index.html
Florida State University	Chemistry and Biochemistry	http://www.chem.fsu.edu/fri2.htm
Georgetown University	Chemistry	http://chemistry.georgetown.edu/people/index.html
Georgia Institute of Technology	Chemistry and Biochemistry	http://www.chemistry.gatech.edu/faculty/index.php
Harvard University	Chemistry and Chemical Biology	http://www.chem.harvard.edu/research/index.php?PHPSESSID=f04ccd7dbae8d1d7b66605501eb5f984
Howard University	Chemistry	http://www.coas.howard.edu/chem/faculty.htm
Indiana University at Bloomington	Chemistry	http://chem.indiana.edu/faculty/
Iowa State University	Chemistry	http://www.chem.iastate.edu/faculty/
Johns Hopkins University	Chemistry	http://chemistry.jhu.edu/interests.html

CHEMISTRY

Name of Institution	Name of Department	URL
Louisiana State University	Chemistry	http://chemistry.lsu.edu/site/People/Faculty/item931.html
Massachusetts Institute of Technology	Chemistry	http://web.mit.edu/chemistry/www/faculty/alpha.html
Michigan State University	Chemistry	http://www.chemistry.msu.edu/Dept_Dir/Fac_Dir.asp
New Mexico State University	Chemistry and Biochemistry	http://www.chemistry.nmsu.edu/Faculty-1.htm
New York University	Chemistry	http://www.nyu.edu/pages/chemistry/faculty/alpha.html
North Carolina State University	Chemistry	http://www.ncsu.edu/chemistry/faculty_list.html
Northwestern University	Chemistry	http://www.chem.northwestern.edu/faculty/
Ohio State University	Chemistry	http://www.chemistry.ohio-state.edu/cgi/directory?Directive=faculty
Oregon State University	Chemistry	http://sci-chem.science.oregonstate.edu/dept_directory.html
Pennsylvania State University	Chemistry	http://www.chem.psu.edu/faculty
Princeton University	Chemistry	http://www.princeton.edu/~chemdept/fss/index.html
Purdue University	Chemistry	http://www.chem.purdue.edu/people/faculty.asp
Rockefeller University	NA	NA
Rutgers, the State University of New Jersey	Chemistry and Chemical Biology	http://rutchem.rutgers.edu/faculty_directory.shtml
Stanford University	Chemistry	http://www.stanford.edu/dept/chemistry/faculty/index.html
State University of New York at Buffalo	Chemistry	http://www.chemistry.buffalo.edu/people/

CHEMISTRY

Name of Institution	Name of Department	URL
State University of New York at Stony Brook	Chemistry	http://www.chem.stonybrook.edu/
Temple University	Chemistry	http://www.temple.edu/chemistry/
Texas A&M University	Chemistry	http://www.chem.tamu.edu/faculty/index.php
Tufts University	Chemistry	http://chem.tufts.edu/
Tulane University	Chemistry	http://www.tulane.edu/~chemstry/faculty.html
University of Alabama at Birmingham	Chemistry	http://www.chem.uab.edu/chemweb/Faculty/faculty.html
University of Arizona	Chemistry	http://www.chem.arizona.edu/faculty/index.php
University of California-Berkeley	Chemistry	http://chem.berkeley.edu/
University of California-Davis	Chemistry	http://www.chem.ucdavis.edu/
University of California-Irvine	Chemistry	http://www.chem.uci.edu/faculty/
University of California-Los Angeles	Chemistry	http://w3.chem.ucla.edu/dir/fsdir.html
University of California-San Diego	Chemistry and Biochemistry	http://www-chem.ucsd.edu/research/research.cfm
University of California-San Francisco	NA	NA
University of California-Santa Barbara	Chemistry and Biochemistry	http://www.chem.ucsb.edu/people/faculty_list.php?flist=go
University of Chicago	Chemistry	http://chemistry.uchicago.edu/faculty.shtml
University of Cincinnati	Chemistry	http://asweb.artsci.uc.edu/CollegeDepts/chemistry/fac_staff/faculty.cfm

CHEMISTRY

Name of Institution	Name of Department	URL
University of Colorado	Chemistry and Biochemistry	http://www.colorado.edu/chemistry/faculty.html
University of Connecticut	Chemistry	http://web.uconn.edu/chemistry/faculty.html
University of Florida	Chemistry	http://web.chem.ufl.edu/people/faculty/
University of Georgia	Chemistry	http://www.chem.uga.edu/phonebook/cgi/directory.cfm?id=1
University of Hawaii at Manoa	Chemistry	http://www.manoa.hawaii.edu/chem/
University of Illinois at Chicago	Chemistry	http://www.chem.uic.edu/WWW/faculty_site/faculty_main.htm
University of Illinois at Urbana-Champaign	Chemistry	http://www.scs.uiuc.edu/chem/faculty/index.html
University of Iowa	Chemistry	http://www.chem.uiowa.edu/
University of Kansas	Chemistry	http://www.chem.ku.edu/faculty/
University of Kentucky	Chemistry	http://www.chem.uky.edu/resources/directory/directory.html
University of Maryland at College Park	Chemistry and Biochemistry	http://www.chem.umd.edu/Faculty_Directory/
University of Massachusetts at Amherst	Chemistry	http://www.chem.umass.edu/Faculty/index.htm
University of Miami	Chemistry	http://www.as.miami.edu/chemistry/people/
University of Michigan	Chemistry	http://www.umich.edu/~michchem/faculty/alpha.html
University of Minnesota, Twin Cities	Chemistry	http://www.chem.umn.edu/directory/faculty_page.lasso
University of Missouri-Columbia	Chemistry	http://www.chem.missouri.edu/faclist.html

CHEMISTRY

Name of Institution	Name of Department	URL
University of Nebraska-Lincoln	Chemistry	http://www.chem.unl.edu/faculty/
University of New Mexico	Chemistry	http://chemistry.unm.edu/faculty.php
University of North Carolina at Chapel Hill	Chemistry	http://www.chem.unc.edu/people/faculty_by_name.html
University of Pennsylvania	Chemistry	http://www.sas.upenn.edu/chem/
University of Pittsburgh	Chemistry	http://www.chem.pitt.edu/p.php?pid=18
University of Rochester	Chemistry	http://www.chem.rochester.edu/Faculty/default.htm
University of Southern California	Chemistry	http://chem.usc.edu/faculty/alpha.html
University of Tennessee, Knoxville	Chemistry	http://www.chem.utk.edu/
University of Texas at Austin	Chemistry and Biochemistry	http://www.cm.utexas.edu/Faculty-and-Research/Faculty-Directory
University of Utah	Chemistry	http://www.chem.utah.edu/people/people_links.html#faculty
University of Virginia	Chemistry	http://www.virginia.edu/chem/people/
University of Washington	Chemistry	http://depts.washington.edu/chemfac/index.html
University of Wisconsin-Madison	Chemistry	http://www.chem.wisc.edu/content/list-people
Utah State University	Chemistry and Biochemistry	http://www.chem.usu.edu/faculty.php
Vanderbilt University	Chemistry	http://www.vanderbilt.edu/AnS/Chemistry/faculty/
Virginia Commonwealth University	Chemistry	http://www.has.vcu.edu/

CHEMISTRY

Name of Institution	Name of Department	URL
Virginia Polytechnic Institute & State University	Chemistry	http://www.chem.vt.edu/index.html
Washington University	Chemistry	http://www.chemistry.wustl.edu/faculty.html
Wayne State University	Chemistry	http://www.chem.wayne.edu/
West Virginia University	Chemistry	http://chemistry.wvu.edu/people
Yale University	Chemistry	http://www.chem.yale.edu/faculty/index.html
Yeshiva University	Biochemistry	http://www.aecom.yu.edu/biochemistry/

CIVIL ENGINEERING

Name of Institution	Name of Department	URL
Arizona State University	Civil and Environmental Engineering	http://www.fulton.asu.edu/fulton/#
Boston University	NA	NA
Brown University	NA	NA
California Institute of Technology	Civil Engineering	http://www.ce.caltech.edu/People/people.html
Carnegie-Mellon University	Civil and Environmental Engineering	http://www.ce.cmu.edu/people/faculty/index.html
Case Western Reserve University	Civil Engineering	http://civil.case.edu/
Colorado State University	Civil Engineering	http://www.engr.colostate.edu/ce/facultystaff/full.shtml
Columbia University	Civil Engineering and Engineering Mechanics	http://www.civil.columbia.edu/
Cornell University	Civil and Environmental Engineering	http://www.cee.cornell.edu/people/
Duke University	Civil and Environmental Engineering	http://www.cee.duke.edu/faculty/
Emory University	NA	NA

CIVIL ENGINEERING

Name of Institution	Name of Department	URL
Florida State University	Civil and Environmental Engineering	http://www.eng.fsu.edu/departments/civil/people/faculty.htm
Georgetown University	NA	NA
Georgia Institute of Technology	Civil and Environmental Engineering	http://www.ce.gatech.edu/fac_staff/faculty-listing/
Harvard University	NA	NA
Howard University	Civil Engineering	http://www.howard.edu/ceacs/departments/CIVIL/Default.htm
Indiana University at Bloomington	NA	NA
Iowa State University	Civil, Construction, and Environmental Engineering	http://www.ccee.iastate.edu/faculty_staff/index.cfm
Johns Hopkins University	Civil Engineering	http://www.civil.jhu.edu/people-departmental-faculty/
Louisiana State University	Civil and Environmental Engineering	http://www.cee.lsu.edu/people/default.aspx
Massachusetts Institute of Technology	Civil and Environmental Engineering	http://cee.mit.edu/faculty
Michigan State University	Civil and Environmental Engineering	http://www.egr.msu.edu/cee/people/faculty.html
New Mexico State University	Civil and Geological	http://cagesun.nmsu.edu/faculty/index.html
New York University	NA	NA
North Carolina State University	Civil, Construction, and Environmental Engineering	http://www.ce.ncsu.edu/faculty/
Northwestern University	Civil and Environmental Engineering	http://www.civil.northwestern.edu/people/index.html#faculty
Ohio State University	Civil, Environmental, and Geodetic	http://www.ceegs.ohio-state.edu/faculty/
Oregon State University	Civil, Construction, and Environmental Engineering	http://ccee.oregonstate.edu/people/faculty/index.html
Pennsylvania State University	Civil and Environmental Engineering	http://www.engr.psu.edu/ce/facstaffdir.html
Princeton University	Civil and Environmental Engineering	http://www.cee.princeton.edu/civfssfac.html

CIVIL ENGINEERING

Name of Institution	Name of Department	URL
Purdue University	Civil Engineering	https://engineering.purdue. edu/CE/People/Faculty/ ViewByAlpha.html
Rockefeller University	NA	NA
Rutgers, the State University of New Jersey	Civil and Environmental Engineering	http://www.civeng.rutgers. edu/faculty/
Stanford University	Civil and Environmental Engineering	http://cee.stanford.edu/about_ faculty/faculty_dir.html
State University of New York at Buffalo	Civil, Structural, and Environment Engineering	http://www.civil.buffalo.edu/ people_fac.shtml
State University of New York at Stony Brook	NA	NA
Temple University	Civil and Environmental Engineering	http://www.temple.edu/ engineering/CEE/directory.html
Texas A&M University	Civil Engineering	https://www.civil.tamu.edu/
Tufts University	Civil and Environmental Engineering	http://engineering.tufts.edu/cee/ people/
Tulane University	Civil and Environmental Engineering	http://www.tulane.edu/%7Ecivil/ faculty.html
University of Alabama at Birmingham	Civil and Environmental Engineering	http://main.uab.edu/soeng/ Templates/Inner.aspx?pid=49363
University of Arizona	Civil Engineering and Engineering Mechanics	http://w3.arizona.edu/~civil/e- files/Phonebook03-04.pdf
University of California- Berkeley	Civil and Environmental Engineering	http://www.ce.berkeley.edu/ faculty/
University of California-Davis	Civil and Environmental Engineering	http://cee.engr.ucdavis.edu/ people/faculty.htm
University of California- Irvine	Civil and Environmental Engineering	http://www.eng.uci.edu/ directory?dept=cee
University of California- Los Angeles	Civil and Environmental Engineering	http://www.cee.ucla.edu/faculty. htm

CIVIL ENGINEERING

Name of Institution	Name of Department	URL
University of California-San Diego	Structural Engineering	http://structures.ucsd.edu/index.php?page=structural_engineering/people/faculty/index
University of California-San Francisco	NA	NA
University of California-Santa Barbara	NA	NA
University of Chicago	NA	NA
University of Cincinnati	Civil and Environmental Engineering	http://www.eng.uc.edu/dept_cee/people/faculty/
University of Colorado	Civil, Environmental, and Architectural Engineering	http://www.colorado.edu/catalog/catalog09-10/engineering/civilengineering.html
University of Connecticut	Civil and Environmental Engineering	http://www.engr.uconn.edu/cee/
University of Florida	Civil and Coastal Engineering	http://www.ce.ufl.edu/people/faculty/division.htm
University of Georgia	NA	NA
University of Hawaii at Manoa	Civil Engineering	http://www.cee.hawaii.edu/
University of Illinois at Chicago	Civil and Material Engineering	http://www.uic.edu/depts/cme/people/faculty/
University of Illinois at Urbana-Champaign	Civil Engineering	http://www.engr.uiuc.edu/faculty/directory.php?dept=civil
University of Iowa	Civil and Environmental Engineering	http://www.cee.engineering.uiowa.edu/people.php
University of Kansas	Civil, Environmental, and Architectural Engineering	http://ceae.engr.ku.edu/
University of Kentucky	Civil Engineering	http://www.engr.uky.edu/ce/faculty_staff/index.html
University of Maryland at College Park	Civil and Environmental Engineering	http://www.umd.edu/directories/
University of Massachusetts at Amherst	Civil and Environmental Engineering	http://www.physics.umass.edu/people/

CIVIL ENGINEERING

Name of Institution	Name of Department	URL
University of Miami	Civil, Architectural, and Environmental Engineering	http://www.cae.miami.edu/
University of Michigan	Civil and Environmental Engineering	http://cee.engin.umich.edu/
University of Minnesota, Twin Cities	Civil Engineering	http://www.ce.umn.edu/people/faculty/
University of Missouri-Columbia	Civil and Environmental Engineering	http://web.missouri.edu/%7Ecivilwww/html/faculty.html - not working
University of Nebraska-Lincoln	Civil Engineering	http://www.civil.unl.edu/faculty/ - not working
University of New Mexico	Civil Engineering	http://www.unm.edu/~civil/faculty.htm - not working
University of North Carolina at Chapel Hill	NA	NA
University of Pennsylvania	NA	NA
University of Pittsburgh	Civil and Environmental Engineering	http://www.engr.pitt.edu/civil/about/faculty/index.html - not working
University of Rochester	NA	NA
University of Southern California	Civil and Environmental Engineering	http://www.usc.edu/dept/civil_eng/dept/
University of Tennessee, Knoxville	Civil and Environmental Engineering	http://www.engr.utk.edu/civil/people/faculty.htm#faculty - not working
University of Texas at Austin	Civil Engineering	http://www.ce.utexas.edu/facultyDir.cfm
University of Utah	Civil and Environmental Engineering	http://www.civil.utah.edu/faculty.htm - not working
University of Virginia	Civil Engineering	http://ce.virginia.edu/people/
University of Washington	Civil and Environmental Engineering	http://www.ce.washington.edu/faculty/faculty.htm - not working
University of Wisconsin-Madison	Civil and Environmental Engineering	http://www.engr.wisc.edu/cee/faculty/
Utah State University	Civil and Environmental Engineering	http://www.engineering.usu.edu/cee/people.html - not working

CIVIL ENGINEERING

Name of Institution	Name of Department	URL
Vanderbilt University	Civil and Environmental Engineering	http://www.cee.vanderbilt.edu/ cee/facstaff.html - not working
Virginia Commonwealth University	NA	NA
Virginia Polytechnic Institute & State University	Civil and Environmental Engineering	http://www.cee.vt.edu/people/ faculty.asp
Washington University	Civil Engineering	http://www.cive.wustl.edu/
Wayne State University	Civil and Environmental Engineering	http://www.eng.wayne.edu/coe/ main.cfm?location=1790
West Virginia University	Civil and Environmental Engineering	http://www.cee.cemr.wvu. edu/faculty/directory. php?type=faculty
Yale University	NA	NA
Yeshiva University	NA	NA

ELECTRICAL ENGINEERING

Name of Institution	Name of Department	URL
Arizona State University	Electrical Engineering	http://ee.fulton.asu.edu/ facultystaff/faculty-directory
Boston University	Electrical and Computer Engineering	http://www.bu.edu/dbin/ece/web/ people/faculty.php
Brown University	NA	NA
California Institute of Technology	Electrical Engineering	http://www.ee2.caltech.edu/ people.html
Carnegie-Mellon University	Electrical and Computer Engineering	http://www.ece.cmu.edu/people/ show.php?type=faculty&range=1
Case Western Reserve University	Electrical Engineering and Computer Science	http://www.eecs.case.edu/people/ faculty/doku.php?id=eecs: people:faculty&section=faculty_ and_staff
Colorado State University	Electrical and Computer Engineering	http://www.engr.colostate. edu/ece/
Columbia University	Electrical Engineering	http://www.ee.columbia.edu/ directory/faculty.html

ELECTRICAL ENGINEERING

Name of Institution	Name of Department	URL
Cornell University	Electrical and Computer Engineering	http://www.ece.cornell.edu/
Duke University	Electrical and Computer Engineering	http://www.ee.duke.edu/People/people.php?type=ALL_FACULTY
Emory University	NA	NA
Florida State University	Electrical and Computer Engineering	http://www.eng.fsu.edu/ece/
Georgetown University	NA	NA
Georgia Institute of Technology	Electrical and Computer Engineering	http://www.ece.gatech.edu/faculty/fac_profiles/ece_faculty.php
Harvard University	Electrical Engineering and Computer Science	http://www.eecs.harvard.edu/index/eecs_faculty.php
Howard University	Electrical and Computer Engineering	http://www.howard.edu/ceacs/Departments/Electrical/Default.htm
Indiana University at Bloomington	NA	NA
Iowa State University	Electrical and Computer Engineering	http://www.ece.iastate.edu/
Johns Hopkins University	Electrical and Computer Engineering	http://www.ece.jhu.edu/FacultyWeb/fac_dir.shtml
Louisiana State University	Electrical and Computer Engineering	http://www.ee.lsu.edu/
Massachusetts Institute of Technology	Electrical Engineering and Computer Science	http://www.eecs.mit.edu/faculty/index.html
Michigan State University	Electrical and Computer Engineering	http://www.egr.msu.edu/ece/fac/#Faculty
New Mexico State University	School of Electrical and Computer Engineering	http://www.ece.nmsu.edu/
New York University	NA	NA
North Carolina State University	Electrical and Computer Engineering	http://www.ece.ncsu.edu/dept/directory/index.php?Group=F&col=&Search=&Images=Yes

ELECTRICAL ENGINEERING

Name of Institution	Name of Department	URL
Northwestern University	Electrical and Computer Engineering	http://www.eecs.northwestern.edu/
Ohio State University	Electrical and Computer Engineering	http://www.ece.osu.edu/
Oregon State University	Electrical Engineering and Computer Science	http://eecs.oregonstate.edu/people/faculty_staff_dir.html
Pennsylvania State University	Electrical Engineering	http://www.ee.psu.edu/faculty/
Princeton University	Electrical Engineering	http://www.ee.princeton.edu/people/faculty-i.php
Purdue University	Electrical and Computer Engineering	https://engineering.purdue.edu/ECE/People/Faculty
Rockefeller University	NA	NA
Rutgers, the State University of New Jersey	Electrical and Computer Engineering	http://www.ece.rutgers.edu/directory/faculty.html
Stanford University	Electrical Engineering	http://www-ee.stanford.edu/faculty.php
State University of New York at Buffalo	Electrical Engineering	http://www.ee.buffalo.edu/
State University of New York at Stony Brook	Electrical and Computer Engineering	http://www.ece.sunysb.edu/
Temple University	Electrical and Computer Engineering	http://www.temple.edu/engineering/ECE/index.html
Texas A&M University	Electrical Engineering	http://ee.tamu.edu/htmlFrames.htm
Tufts University	Electrical and Computer Engineering	http://www.ece.tufts.edu/people/faculty/
Tulane University	Electrical Engineering and Computer Science	http://www.sse.tulane.edu/pages/home.php
University of Alabama at Birmingham	Electrical and Computer Engineering	http://main.uab.edu/soeng/Templates/Inner.aspx?pid=49364
University of Arizona	Electrical and Computer Engineering	http://ece.arizona.edu/faculty.php

ELECTRICAL ENGINEERING

Name of Institution	Name of Department	URL
University of California-Berkeley	Electrical Engineering and Computer Sciences	http://www.eecs.berkeley.edu/Faculty/Lists/gallery.shtml
University of California-Davis	Electrical and Computer Engineering	http://www.ece.ucdavis.edu/people/departmentalfaculty.html
University of California-Irvine	Electrical Engineering and Computer Science	http://www.eng.uci.edu/directory?dept=eecs
University of California-Los Angeles	Electrical Engineering	http://www.ee.ucla.edu/Directory-home.htm
University of California-San Diego	Electrical and Computer Engineering	http://www.ece.ucsd.edu/faculty_research/
University of California-San Francisco	NA	NA
University of California-Santa Barbara	Electrical and Computer Engineering	http://www.ece.ucsb.edu/directory.shtml
University of Chicago	NA	NA
University of Cincinnati	Electrical and Computer Engineering and Computer Science	http://www.uc.edu/cas/ecet/
University of Colorado	Electrical and Computer Engineering	http://ecee.colorado.edu/
University of Connecticut	Electrical and Computer Engineering	http://www.engr.uconn.edu/ece/index.php
University of Florida	Electrical and Computer Engineering	http://www.ece.ufl.edu/people/directories/main.html
University of Georgia	NA	NA
University of Hawaii at Manoa	Electrical Engineering	http://www-ee.eng.hawaii.edu/faculty/
University of Illinois at Chicago	Electrical and Computer Engineering	http://www.ece.uic.edu/Faculty/Faculty.html
University of Illinois at Urbana-Champaign	Electrical and Computer Engineering	http://www.ece.uiuc.edu/faculty/facalpha.asp
University of Iowa	Electrical and Computer Engineering	http://www.engineering.uiowa.edu/ece/faculty.htm

ELECTRICAL ENGINEERING

Name of Institution	Name of Department	URL
University of Kansas	Electrical Engineering and Computer Science	http://www.eecs.ku.edu/people
University of Kentucky	Electrical and Computer Engineering	http://www.engr.uky.edu/ece/faculty_staff/index.html
University of Maryland at College Park	Electrical and Computer Engineering	http://www.ee.umd.edu/
University of Massachusetts at Amherst	Electrical and Computer Engineering	http://www.ecs.umass.edu/index.pl?id=4425&isa=Category&op=show
University of Miami	Electrical and Computer Engineering	http://www.ece.miami.edu/
University of Michigan	Electrical Engineering and Computer Science	https://www.eecs.umich.edu/eecs/faculty/faculty.html
University of Minnesota, Twin Cities	Electrical and Computer Engineering	http://www.ece.umn.edu/
University of Missouri-Columbia	Electrical and Computer Engineering	http://engineering.missouri.edu/ece/faculty-staff/
University of Nebraska-Lincoln	Electrical Engineering	http://engineering.unl.edu/academicunits/ElectricalEngineering/faculty-staff/index.shtml
University of New Mexico	Electrical and Computer Engineering	http://www.ece.unm.edu/professors/professors.htm
University of North Carolina at Chapel Hill	NA	NA
University of Pennsylvania	Electrical and Systems Engineering	http://www.ese.upenn.edu/people/faculty.html
University of Pittsburgh	Electrical and Computer Engineering	http://www.engr.pitt.edu/electrical/research/areas.html
University of Rochester	Electrical and Computer Engineering	http://www.ece.rochester.edu/html/people/faculty.html
University of Southern California	Electrical Engineering	http://ee.usc.edu/faculty_staff/faculty_directory/
University of Tennessee, Knoxville	Electrical and Computer Engineering	http://www.eecs.utk.edu/people/faculty/main?redirect=1

ELECTRICAL ENGINEERING

Name of Institution	Name of Department	URL
University of Texas at Austin	Electrical and Computer Engineering	http://www.ece.utexas.edu/faculty/directory/
University of Utah	Electrical and Computer Engineering	http://www.cce.utah.edu/departments/ece.php
University of Virginia	Electrical and Computer Engineering	http://www.ee.virginia.edu/people.php
University of Washington	Electrical Engineering	http://www.ee.washington.edu/people/faculty/
University of Wisconsin-Madison	Electrical and Computer Engineering	http://www.engr.wisc.edu/ece/faculty/
Utah State University	Electrical and Computer Engineering	http://www.engineering.usu.edu/ece/faculty_staff/
Vanderbilt University	Electrical Engineering and Computer Science	
Virginia Commonwealth University	NA	NA
Virginia Polytechnic Institute & State Univesity	Electrical and Computer Engineering	http://www.ecpe.vt.edu/faculty/index.html
Washington University	Electrical and Systems Engineering	http://www.ese.wustl.edu/About/People.asp?Filter=FacStaff
Wayne State University	Electrical and Computer Engineering	http://www.eng.wayne.edu/page.php?id=199
West Virginia University	Computer Science and Electrical Engineering	http://www.lcsee.cemr.wvu.edu/faculty/directory.php?type=faculty
Yale University	Electrical Engineering	http://www.seas.yale.edu/faculty.php#/?department=4
Yeshiva University	NA	NA

MATHEMATICS

Name of Institution	Name of Department	URL
Arizona State University	Mathematics and Statistics	http://math.asu.edu/people/faculty.html
Boston University	Mathematics	http://math.bu.edu/people/faculty.html

MATHEMATICS

Name of Institution	Name of Department	URL
Brown University	Mathematics	http://www.math.brown.edu/faculty/faculty.html
California Institute of Technology	Mathematics	http://www.math.caltech.edu/people/profs.html
Carnegie-Mellon University	Mathematics	http://www.math.cmu.edu/math/faculty.html
Case Western Reserve University	Mathematics	http://www.case.edu/artsci/math/
Colorado State University	Mathematics	http://www.math.colostate.edu/faculty/facultylist.html#currfac
Columbia University	Mathematics	http://www.math.columbia.edu/lists/faculty-alpha.html
Cornell University	Mathematics	http://www.math.cornell.edu/People/faculty.html
Duke University	Mathematics	http://fds.duke.edu/db/aas/math/faculty/
Emory University	Mathematics and Computer Science	http://www.mathcs.emory.edu/
Florida State University	Mathematics	http://www.math.fsu.edu/People/faculty.php
Georgetown University	Mathematics	http://math.georgetown.edu/
Georgia Institute of Technology	Mathematics	http://www.math.gatech.edu/people/faculty/
Harvard University	Mathematics	http://www.math.harvard.edu/people/all.html
Howard University	Mathematics	http://www.coas.howard.edu/mathematics/faculty.html
Indiana University at Bloomington	Mathematics	http://www.math.indiana.edu/people/faculty.phtml
Iowa State University	Mathematics	http://www.math.iastate.edu/About/aboutFaculty.html
Johns Hopkins University	Mathematics	http://www.mathematics.jhu.edu/new/people/people-faculty.htm
Louisiana State University	Mathematics	http://www.math.lsu.edu/

MATHEMATICS

Name of Institution	Name of Department	URL
Massachusetts Institute of Technology	Mathematics	http://math.mit.edu/people/listing.php
Michigan State University	Mathematical Sciences	http://mathdata.msu.edu/FD/RW/F1.html
New Mexico State University	Mathematics	http://www.math.nmsu.edu/
New York University	Mathematics	http://www.math.nyu.edu/people/#faculty
North Carolina State University	Mathematics	http://www.math.ncsu.edu/people/index.php
Northwestern University	Mathematics	http://www.math.northwestern.edu/people/faculty.html
Ohio State University	Mathematics	http://www.math.ohio-state.edu/contacts
Oregon State University	Mathematics	http://www.math.oregonstate.edu/people/list/faculty
Pennsylvania State University	Mathematics	http://www.math.psu.edu/
Princeton University	Mathematics	http://www.math.princeton.edu/menusa/index4.html
Purdue University	Mathematics	http://www.math.purdue.edu/people/faculty
Rockefeller University	NA	NA
Rutgers, the State University of New Jersey	Mathematics	http://www.math.rutgers.edu/people/faculty.html#fac
Stanford University	Mathematics	http://math.stanford.edu/directory/faculty.html
State University of New York at Buffalo	Mathematics	http://www.math.buffalo.edu/fac_staff_list.html
State University of New York at Stony Brook	Mathematics	http://www.math.sunysb.edu/html/faculty-alph.shtml
Temple University	Mathematics	http://math.temple.edu/
Texas A&M University	Mathematics	http://www.math.tamu.edu/directory/faculty.php

MATHEMATICS

Name of Institution	Name of Department	URL
Tufts University	Mathematics	http://www.tufts.edu/as/math/facultystaff.html
Tulane University	Mathematics	http://www.math.tulane.edu/faculty/faculty.html
University of Alabama at Birmingham	Mathematics	http://www.math.uab.edu/people.html
University of Arizona	Mathematics	http://math.arizona.edu/people/categorical.html?category=fac
University of California-Berkeley	Mathematics	http://math.berkeley.edu/people_faculty.html
University of California-Davis	Mathematics	http://www.math.ucdavis.edu/people
University of California-Irvine	Mathematics	http://math.uci.edu/personnel/faculty.html
University of California-Los Angeles	Mathematics	http://www.math.ucla.edu/people/faculty/
University of California-San Diego	Mathematics	http://math.ucsd.edu/people/directory.pl?category=faculty_by_name
University of California-San Francisco	NA	NA
University of California-Santa Barbara	Mathematics	http://www.math.ucsb.edu/department/faculty.php
University of Chicago	Mathematics	http://www.math.uchicago.edu/people/
University of Cincinnati	Mathematical Sciences	http://asweb.artsci.uc.edu/collegedepts/math/facStaff/index.aspx
University of Colorado	Mathematics	http://www.colorado.edu/math/people/index.html
University of Connecticut	Mathematics	http://www.math.uconn.edu/index.php?content=People/index&title=People
University of Florida	Mathematics	http://www.math.ufl.edu/people.html

MATHEMATICS

Name of Institution	Name of Department	URL
University of Georgia	Mathematics	http://www.math.uga.edu/dept_members/professors.html
University of Hawaii at Manoa	Mathematics	http://www.math.hawaii.edu/
University of Illinois at Chicago	Mathematics	http://www.math.uic.edu/people/faculty
University of Illinois at Urbana-Champaign	Mathematics	http://www.math.uiuc.edu/People/members.html
University of Iowa	Mathematics	http://www.math.uiowa.edu/directory/faculty.shtml
University of Kansas	Mathematics	http://www.math.ku.edu/people/faculty.html
University of Kentucky	Mathematics	http://www.ms.uky.edu/~math/Info/03-facc.htm
University of Maryland at College Park	Mathematics	http://www.math.umd.edu/directory/
University of Massachusetts at Amherst	Mathematics and Statistics	http://www.math.umass.edu/Directory/faculty.html
University of Miami	Mathematics	http://www.math.miami.edu/dire/faculty.htm
University of Michigan	Mathematics	http://www.math.lsa.umich.edu/people/faculty.shtml
University of Minnesota, Twin Cities	Mathematics	http://www.math.umn.edu/arb/faculty.shtml
University of Missouri-Columbia	Mathematics	http://www.math.missouri.edu/personnel/faculty.html
University of Nebraska-Lincoln	Mathematics	http://www.math.unl.edu/people/faculty/
University of New Mexico	Mathematics and Statistics	http://www.math.unm.edu/internal/facultyList.php
University of North Carolina at Chapel Hill	Mathematics	http://www.math.unc.edu/people/faculty/
University of Pennsylvania	Mathematics	http://www.math.upenn.edu/FacData.html

MATHEMATICS

Name of Institution	Name of Department	URL
University of Pittsburgh	Mathematics	http://www.math.pitt.edu/people.html
University of Rochester	Mathematics	http://www.math.rochester.edu/people/faculty/
University of Southern California	Mathematics	http://math.usc.edu/mathematics/people/faculty/
University of Tennessee, Knoxville	Mathematics	http://www.math.utk.edu/People/
University of Texas at Austin	Mathematics	http://www.ma.utexas.edu/cgi-bin/addtab/facabc.html
University of Utah	Mathematics	http://www.math.utah.edu/people/faculty.html
University of Virginia	Mathematics	http://www.math.virginia.edu/directory.htm#FACULTY
University of Washington	Mathematics	http://www.math.washington.edu/People/fac_contact.php
University of Wisconsin-Madison	Mathematics	http://www.math.wisc.edu/~apache/psdbfaculty.html
Utah State University	Mathematics and Statistics	http://www.math.usu.edu/
Vanderbilt University	Mathematics	http://www.math.vanderbilt.edu/
Virginia Commonwealth University	Mathematics	http://www.math.vcu.edu/faculty.html
Virginia Polytechnic Institute & State University	Mathematics	http://www.math.vt.edu/people.php?content=list&type=Faculty
Washington University	Mathematics	http://www.math.wustl.edu/
Wayne State University	Mathematics	http://www.math.wayne.edu/addr.html#faculty
West Virginia University	Mathematics	http://www.math.wvu.edu/ldap_search.php?title=*professor
Yale University	Mathematics	http://www.yale.edu/ycpo/ycps/M-P/mathFM.html
Yeshiva University	NA (undergrad only)	NA

PHYSICS AND ASTRONOMY

Name of Institution	Name of Department	URL
Arizona State University	Physics and Astronomy	http://physics.asu.edu/people/faculty
Boston University	Physics	http://physics.bu.edu/people/by_type/1
Brown University	Physics	http://www.physics.brown.edu/people/
California Institute of Technology	Physics	http://www.pma.caltech.edu/GSR/facresearch.html
Carnegie-Mellon University	Physics	http://info.phys.cmu.edu/people/fac.asp
Case Western Reserve University	Physics	http://www.phys.cwru.edu/faculty/
Colorado State University	Physics	http://www.physics.colostate.edu/People/faculty_html
Columbia University	Physics	http://www.columbia.edu/cu/physics/index.html
Cornell University	Physics	http://www.physics.cornell.edu/people/
Duke University	Physics	http://fds.duke.edu/db/aas/Physics/faculty/
Emory University	Physics	http://www.physics.emory.edu/faculty/
Florida State University	Physics	http://www.physics.fsu.edu/
Georgetown University	Physics	http://magus.physics.georgetown.edu/faculty.htm
Georgia Institute of Technology	Physics	http://www.physics.gatech.edu/people/professors.html
Harvard University	Physics	http://physics.harvard.edu/faculty.htm
Howard University	Physics and Astronomy	http://www.physics1.howard.edu/
Indiana University at Bloomington	Physics	http://www.iub.edu/~iubphys/
Iowa State University	Physics and Astronomy	http://www.physics.iastate.edu/
Johns Hopkins University	Physics and Astronomy	http://physics-astronomy.jhu.edu/people/faculty/

PHYSICS AND ASTRONOMY

Name of Institution	Name of Department	URL
Louisiana State University	Physics and Astronomy	http://www.phys.lsu.edu/newwebsite/people/index.html
Massachusetts Institute of Technology	Physics	http://web.mit.edu/physics/facultyandstaff/faculty/faculty_alpha_listing.html
Michigan State University	Physics and Astronomy	http://extranet.pa.msu.edu/directory/dirsearch.asp?GroupID=ALL&lastname=&classID=fac&submit=Perform+Search
New Mexico State University	Physics	http://physics.nmsu.edu/Physics/people/faculty.html
New York University	Physics	http://www.physics.nyu.edu/people/faculty.html
North Carolina State University	Physics	http://www.physics.ncsu.edu/people/index.html
Northwestern University	Physics and Astronomy	http://www.physics.northwestern.edu/
Ohio State University	Physics	http://www.physics.ohio-state.edu/directory/directory_report.php?faculty=1
Oregon State University	Physics	http://physics.orst.edu/People-faculty
Pennsylvania State University	Physics	http://www.phys.psu.edu/people/faculty/
Princeton University	Physics	http://www.princeton.edu/physics/people/
Purdue University	Physics	http://www.physics.purdue.edu/people/index.shtml
Rockefeller University	NA	NA
Rutgers, the State University of New Jersey	Physics and Astronomy	http://physcgi.rutgers.edu/~physdir/directory-lists.pl?faculty
Stanford University	Physics	http://www.stanford.edu/dept/physics/people/faculty.html
State University of New York at Buffalo	Physics	http://electron.physics.buffalo.edu/facultystaff.html

PHYSICS AND ASTRONOMY

Name of Institution	Name of Department	URL
State University of New York at Stony Brook	Physics and Astronomy	http://insti.physics.sunysb.edu/Physics/faculty.shtml
Temple University	Physics	http://www.temple.edu/physics/directory/faculty/
Texas A&M University	Physics	http://www.physics.tamu.edu/people/showgroup.php?group=faculty
Tufts University	Physics and Astronomy	http://ase.tufts.edu/physics/faculty.htm
Tulane University	Physics	http://www.physics.tulane.edu/
University of Alabama at Birmingham	Physics	http://www.phy.uab.edu/
University of Arizona	Physics	http://www.physics.arizona.edu/physics2006/people.php?page=faculty
University of California-Berkeley	Physics	http://physics.berkeley.edu/index.php?option=com_dept_management&Itemid=312
University of California-Davis	Physics	http://www.physics.ucdavis.edu/facultylist.php
University of California-Irvine	Physics and Astronomy	http://www.physics.uci.edu/NEW/faculty.shtml
University of California-Los Angeles	Physics and Astronomy	http://personnel.physics.ucla.edu/directory/index.php
University of California-San Diego	Physics	http://www-physics.ucsd.edu/fac_staff/profiles/index.shtml
University of California-San Francisco	NA	NA
University of California-Santa Barbara	Physics	http://www.physics.ucsb.edu/People/List.php3?F
University of Chicago	Physics	http://physics.uchicago.edu/faculty/

PHYSICS AND ASTRONOMY

Name of Institution	Name of Department	URL
University of Cincinnati	Physics	http://homepages.uc.edu/physics/facultyStaff/index.html
University of Colorado	Physics	http://www.colorado.edu/physics/Web/directory/index.html
University of Connecticut	Physics	http://www.physics.uconn.edu/
University of Florida	Physics	http://www.phys.ufl.edu/faculty/
University of Georgia	Physics and Astronomy	http://www.physast.uga.edu/people.html
University of Hawaii at Manoa	Physics and Astronomy	http://www.phys.hawaii.edu/
University of Illinois at Chicago	Physics	http://www.uic.edu/casp/depts/phys/people/index.asp
University of Illinois at Urbana-Champaign	Physics	http://www.uic.edu/casp/depts/phys/people/index.asp
University of Iowa	Physics and Astronomy	http://www.physics.uiowa.edu/faculty/
University of Kansas	Physics and Astronomy	http://www.physics.ku.edu/faculty/
University of Kentucky	Physics and Astronomy	http://www.pa.uky.edu/fac_roster.html
University of Maryland at College Park	Physics	http://www.physics.umd.edu/people/faculty/
University of Massachusetts at Amherst	Physics	http://www.physics.umass.edu/directory/
University of Miami	Physics	http://web.physics.miami.edu/
University of Michigan	Physics	http://www.lsa.umich.edu/physics/people/faculty
University of Minnesota, Twin Cities	Physics	http://www.physics.umn.edu/directory/faculty.html
University of Missouri-Columbia	Physics and Astronomy	http://www.physics.missouri.edu/people/faculty.shtml
University of Nebraska-Lincoln	Physics and Astronomy	http://www.unl.edu/ucomm/facstaff/

PHYSICS AND ASTRONOMY

Name of Institution	Name of Department	URL
University of New Mexico	Physics and Astronomy	http://panda.unm.edu/people/faculty_listing.html
University of North Carolina at Chapel Hill	Physics and Astronomy	http://www.physics.unc.edu/directory/directory.php?section=1&mode=text&param=99
University of Pennsylvania	Physics and Astronomy	http://www.physics.upenn.edu/people/faculty.html
University of Pittsburgh	Physics and Astronomy	http://www.phyast.pitt.edu/people/faculty.php
University of Rochester	Physics and Astronomy	http://www.pas.rochester.edu/
University of Southern California	Physics	http://physics.usc.edu/Faculty/
University of Tennessee, Knoxville	Physics and Astronomy	http://www.phys.utk.edu/people_faculty.html
University of Texas at Austin	Physics	http://www.ph.utexas.edu/faculty-list.html
University of Utah	Physics	http://www.physics.utah.edu/people/faculty.html
University of Virginia	Physics	http://www.phys.virginia.edu/People/Faculty-list.asp?CLASS=Faculty&SUBCLASS=Faculty
University of Washington	Physics	http://www.phys.washington.edu/
University of Wisconsin-Madison	Physics	http://www.physics.wisc.edu/people/directory.html
Utah State University	Physics	http://www.physics.usu.edu/department/faculty/
Vanderbilt University	Physics and Astronomy	http://www.physics.vanderbilt.edu/
Virginia Commonwealth University	Physics	http://www.has.vcu.edu/phy/
Virginia Polytechnic Institute & State University	Physics	http://www.phys.vt.edu/people_page.HTML
Washington University	Physics	http://www.physics.wustl.edu/

PHYSICS AND ASTRONOMY

Name of Institution	Name of Department	URL
Wayne State University	Physics and Astronomy	http://www.physics.wayne.edu//people/faculty.html
West Virginia University	Physics	http://physics.wvu.edu/people
Yale University	Physics	http://www.yale.edu/physics/contact_us/directory.shtml
Yeshiva University	NA (undergraduate only)	NA

Appendix 2-1
Review of Literature and Relevant Research

PROFILE OF WOMEN IN ACADEMIC
SCIENCE AND ENGINEERING: 1995-2003

The 2001 National Academies' study, *From Scarcity to Visibility: Gender Differences in the Careers of Doctoral Scientists and Engineers* (NRC, 2001a), examined the careers of men and women scientists and engineers using data from the Survey of Doctorate Recipients (SDR) for four selected years: 1973, 1979, 1989, and 1995. The first part of this appendix provides descriptive data from the SDR for 1995 to 2003, the time period when the surveys of faculty and departments were initiated.[1] This overview presents data on basic trends in female participation and standing among science and engineering (S&E) faculty for 1995 to 2003, including the number of employed doctorates, the fields in which scientists and engineers worked, and the proportions who worked in academia. Academics are further disaggregated by the types of institutions in which they worked, their fields of study, their tenure status, and their professorial rank. It is important to remember that the SDR covers doctoral recipients in all fields of science and engineering and working in all sectors of the economy. This appendix focuses only on those doctoral scientists and engineers who were employed full-time and whose doctorate was in the natural sciences and engineering, excluding the social sciences.[2]

THE DOCTORAL POOL

The number and percentage of women receiving doctorates in S&E grew from 8,648 (31.7 percent) in 1996 to 10,533 (37.7 percent) in 2005, as shown in Figure A2-1.

Increases in women's participation differed by field. Growth was particularly evident, as noted in Table A2-1, in civil engineering, the agricultural sciences, and the earth, atmospheric, and oceanic sciences. But every field, other than industrial/manufacturing engineering, saw increases in the proportion of doctorates awarded to women over the 10-year period.[3]

---

[1] The results of analyses are not strictly comparable, as the earlier report used a different definition of S&E, among other differences.

[2] From the Survey of Earned Doctorates (SED) field list, this is equivalent to any field coded from 005 to 599.

[3] The one recent exception appears to be the medical or health sciences, where the proportion of women among Ph.D.s seemed to have leveled off.

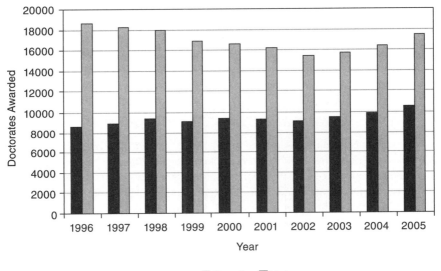

FIGURE A2-1 Number of doctorates awarded annually in science and engineering by gender, 1996-2005.
NOTE: These data are for all science and engineering fields, including the social and behavioral sciences.
SOURCE: Hill (2006). Adapted from Tables 2 and 3.

EMPLOYMENT STATUS

In 2003, the National Science Foundation (NSF) identified 492,440 doctoral scientists and engineers (or 685,300 if the social sciences and psychology are included) (NSF, 2006). Most of these doctoral scientists and engineers worked full-time. However, women were slightly less likely to be employed full-time.

In a previous analysis of SDR data, the National Research Council (NRC) (2001a:64) found "after completion of the doctorate, a greater proportion of women than men do not attain full-time careers in science and engineering." For example, in 1973, 91 percent of male scientists and engineers were working full-time, compared to 71 percent of females. By 1995, this 20 percent gap had been reduced to around 10 percent—partly because the percentage of men working full-time dropped.[4] For all years surveyed, women were more likely than men to be not working and not seeking work, or working part-time. For most years examined, women were more likely than men to be not working, but seeking work. About

---

[4] Recall that Long's definition of S&E includes the social and behavioral sciences and is thus broader than the definition employed here.

TABLE A2-1 Percentage of Women Among Science And Engineering Doctorates, 1996 and 2005

Field	1996	2005	2005–1996
Science and engineering	31.7	37.7	6.0
Science	37.6	43.4	5.8
Agricultural sciences	27.2	36.2	9.0
Biological sciences	42.2	48.8	6.6
Computer sciences	15.1	19.8	4.7
Earth, atmospheric, and oceanic sciences	21.0	34.1	13.1
Mathematics	20.6	27.1	6.5
Physical sciences	21.9	26.7	4.8
Astronomy	21.4	26.3	4.9
Chemistry	28.2	34.0	5.8
Physics	13.0	15.0	2.0
Psychology	66.7	68.0	1.3
Social sciences	36.5	44.7	8.2
Engineering	12.3	18.3	6.0
Aeronautical/astronautical engineering	8.4	13.2	4.8
Chemical engineering	17.9	24.0	6.1
Civil engineering	11.3	23.2	11.9
Electrical engineering	9.7	13.4	3.7
Industrial/manufacturing engineering	19.7	18.5	−1.2
Materials/metallurgical engineering	14.6	22.2	7.6
Mechanical engineering	7.4	12.3	4.9
Other engineering	16.6	23.3	7.2

SOURCE: Hill (2006). Adapted from Table 3.

4 percent of female S&E doctorates were not working and not seeking work. These were fully trained doctorates who were not working in S&E.[5]

"Employment status" consisted of four mutually exclusive categories: employed full-time, employed part-time, unemployed but seeking work, and unemployed and not seeking work. Figure A2-2 examines full-time employment and compares the percentages of full-time employed doctoral scientists and engineers[6] to the total number of doctoral scientists and engineers. As this figure shows, women were less likely to be employed full-time than men, although the rate for both men and women was dropping slightly over time, and the gap was closing.

---

[5] The committee's charge did not include a focus on exploring the reasons for gender differences in labor force outcomes outside of academia. Readers should refer to Long (2001) and Xie and Shauman (2003) for a discussion of such factors.

[6] These data are for just the natural sciences and engineering.

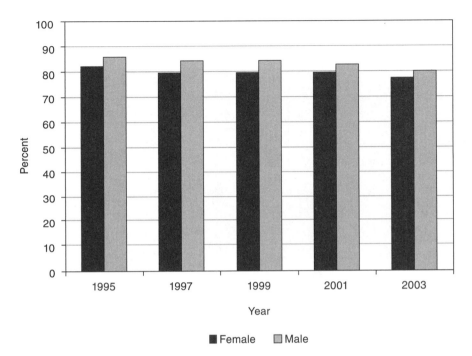

FIGURE A2-2 Percentage of all doctoral scientists and engineers who were employed full-time by gender, 1995-2003.
SOURCE: National Science Foundation, Survey of Doctorate Recipients, 1995-2003. Tabulated by the NRC.

This finding was consistent with the earlier work of NRC (2001a) and others, who employed different analyses. For example, the NSF (WMPDSE, 2002) noted "women with either an S&E degree or in an S&E occupation are less likely than men to be in the labor force (that is, either employed or seeking employment). Among those in the labor force, women were more likely than men to be unemployed." The NSF also noted:

> A higher percentage of women than men with either an S&E degree or in an S&E occupation are employed part time. Of those who were employed in 1999, 19 percent of women and 6 percent of men were employed part-time. Women who are employed part-time are less likely than men to prefer full-time employment. Also, women who are employed part-time are far more likely than men to cite family responsibilities as the reason for their employment status: 48 percent of the women working part-time and 12 percent of the men cited family

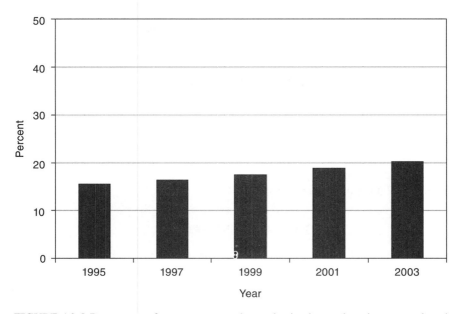

FIGURE A2-3 Percentage of women among doctoral scientists and engineers employed full-time, 1995-2003.
SOURCE: National Science Foundation, Survey of Doctorate Recipients, 1995-2003. Tabulated by the NRC.

responsibilities as the reason for their work status in 1999. On the other hand, 41 percent of men and 8 percent of women cited retirement as the reason for part-time employment. Thus, as with unemployment, variations in male/female age distribution, as well as varying family responsibilities, are factors in part-time employment choices.[7]

Figure A2-3 examines the proportion of women among full-time employed doctoral scientists and engineers between 1995 and 2003. The proportion of women among those employed full-time, while still small, was rising slowly. Increases "in the number of women among new Ph.D.s do not translate directly into increases in the proportion of women in the science and engineering labor force. Each new cohort of Ph.D.s represents only a small fraction of the total number of scientists and engineers. The proportion of women in the S&E labor force must increase slowly as older, predominantly male cohorts retire and are replaced by new cohorts that have a greater proportion of women" (NRC, 2001a:63).

---

[7] Ellipses omitted.

EMPLOYMENT DISCIPLINE

This section briefly examines the distribution of doctoral scientists and engineers employed full-time by field and gender. As shown in Figure A2-4, women employed in the biological, physical, and health sciences were the most likely to be working full-time. In the case of men, those who were employed in engineering and the physical sciences were more likely to be working full-time.

Figure A2-5 examines the percentage of women among doctorates employed full-time in six different disciplines. Although the percentage of women among scientists and engineers was rising, women still made up a small fraction of those employed in the agricultural sciences, engineering, mathematics and computer sciences, and the physical sciences.

EMPLOYMENT SECTOR

This section considers the employment sector of those who were employed full-time. NRC (2001a:102) noted that "sector of employment is a fundamental dimension of the scientific career that affects work experience, opportunities,

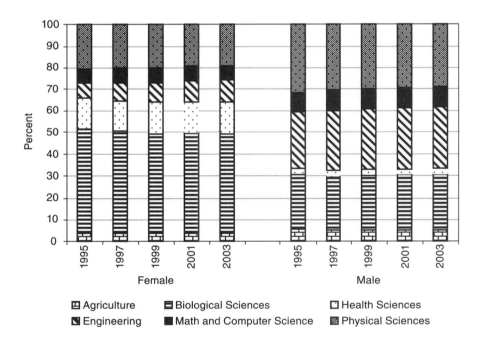

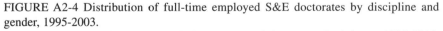

FIGURE A2-4 Distribution of full-time employed S&E doctorates by discipline and gender, 1995-2003.
SOURCE: National Science Foundation, Survey of Doctorate Recipients, 1995-2003. Tabulated by the NRC.

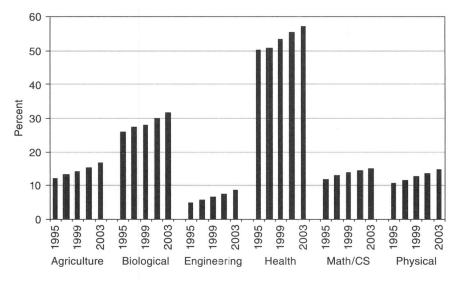

FIGURE A2-5 Percentage of females among doctorates employed full-time by discipline, 1995-2003.
SOURCE: National Science Foundation, Survey of Doctorate Recipients, 1995-2003. Tabulated by the NRC.

employment security, and prestige." An often-used distinction among employment sectors for doctorate holders in S&E is industry, government, and education. *Often, education is narrowly defined to encompass doctoral scientists and engineers working at colleges and universities that award at least a two-year degree (NRC, 2001a). In this section, however, education includes K-12.* Outside of education, the other employment sectors include industry not-for-profit organizations; self-employed persons; local, state, or federal government; or the U.S. military.

According to previous literature, employed women with doctorates in S&E were more likely to be in academia and less likely to be in industry (NRC, 2001a). This finding was echoed by the NSF, which noted that women were more likely than men to be at 4-year academic institutions and less likely to be in business or industry (NSF, 2007). The authors argued that these differences "primarily stem from differences in occupation. Women are less likely than men to be engineers or physical scientists, which are occupations that tend to be in business or industry" (p. 66). The NSF's final point, as well as findings from NRC (2001a), suggested that differences in employment sector vary by discipline; that is, men and women in different areas of S&E distribute themselves differently across possible employment sectors.

Table A2-2 and Figure A2-6 examine the distribution of male and female S&E doctorates employed full-time across two employment sectors: Education

TABLE A2-2 Doctoral Scientists and Engineers Employed Full-Time by Sector and Gender, 1995-2003

Gender/Sector	Years				
	1995	1997	1999	2001	2003
Men					
Education	124,770	125,252	128,335	128,170	131,628
Other	151,115	163,076	179,519	184,260	179,588
Percent Education	.45	.43	.42	.41	.42
Women					
Education	29,759	32,659	35,726	39,621	43,828
Other	21,195	24,126	29,880	33,585	36,117
Percent Education	.58	.58	.54	.54	.55

SOURCE: National Science Foundation, Survey of Doctorate Recipients, 1995-2003. Tabulated by the NRC.

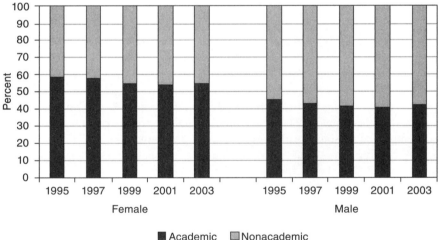

FIGURE A2-6 Percentage of doctorates employed full-time in education and other sectors by gender, 1995-2003.
NOTE: The percentage of women employed full-time in the education sector appeared to be increasing (see Figure A2-7).
SOURCE: National Science Foundation, Survey of Doctorate Recipients, 1995-2003. Tabulated by the NRC.

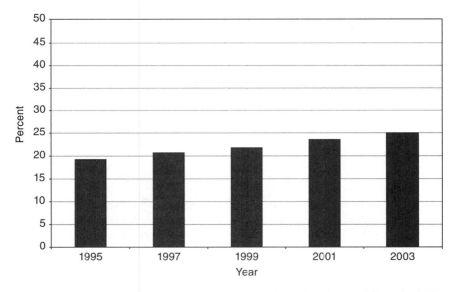

FIGURE A2-7 Percentage of women among the full-time education workforce, including K-12 education, 1995-2003.
SOURCE: National Science Foundation, Survey of Doctorate Recipients, 1995-2003. Tabulated by the NRC.

and Other.[8] As Figure A2-6 shows, women were more likely to be in the education sector than men.

ACADEMICS

Male and female academics can be categorized along several dimensions. The first section examines academics by field and by the type of higher education institution in which they worked, followed by the distribution of male and female faculty across tenure status and rank. The term "academic" is used here to denote *faculty, which are personnel with teaching or research duties, who are employed at a higher education institution (college or university), and who are further identified as being tenured or on tenure track or as holding the rank of assistant, associate, or full professor.*

---

[8] Other includes industry, government, and the nonprofit sector. Education in this table includes K-12 positions.

Distribution by Discipline

As Figure A2-8 shows, more than half of the faculty in the health sciences in 1995-2003 were women. The biological sciences also had relatively large proportions of female faculty (20-30 percent). In the other four disciplines, and especially in engineering, women made up a small fraction of the faculty. In every field, however, the proportion of women among faculty was smaller than the corresponding proportion of women among those earning a doctorate in the discipline.

Distribution by Institution Type

In this section, we focus on doctoral scientists and engineers who were employed at Research I institutions, consisting of institutions that "offer a full range of baccalaureate programs, are committed to graduate education through the doctorate degree, and give high priority to research. They award 50 or more doctoral degrees each year. In addition, they receive annually at least $40 million or more in federal support." Using the 1994 Carnegie classification, there are 89 Research I institutions in the United States.[9] The following tables group employed doctoral scientists and engineers by the institutional category they reported in the SDR. There are seven possible institutional categories: Research I, Research II, doctoral-granting, master's-granting, medical colleges, baccalaureate (4-year institutions), and other (including 2-year institutions). None of the categories overlaps.

As Figure A2-9 shows, the highest percentage of female faculty was found in medical colleges, and the lowest percentage of women was found at Research II institutions. Among the other types of institutions, women tended to make up between 20 and 25 percent of S&E faculty. The percentage of female faculty employed at Research I institutions was growing steadily in 1995-2003.

Tenure Status

How likely were women to be granted tenure? Using the SDR, we examined tenure status by gender by comparing faculty with tenure to faculty who were untenured but on the tenure-track, considering each academic discipline separately. As Figure A2-10 shows, the percentage of women among tenured faculty appeared to be growing in 1995-2003 in all fields, while the percentage of women among tenure-track faculty was growing in some fields, including engineering.

---

[9] See Alexander C. McCormick, "The 2000 Carnegie Classification: Background and Description (excerpt)," available at http://www.carnegiefoundation.org/dynamic/downloads/file_1_341.pdf [accessed on November 4, 2008]. The Carnegie Foundation updated its classification system in 2005 and is available at http://www.carnegiefoundation.org/classifications/.

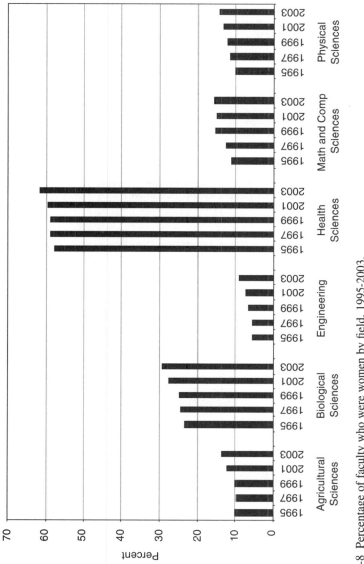

FIGURE A2-8 Percentage of faculty who were women by field, 1995-2003.
SOURCE: National Science Foundation, Survey of Doctorate Recipients, 1995-2003.

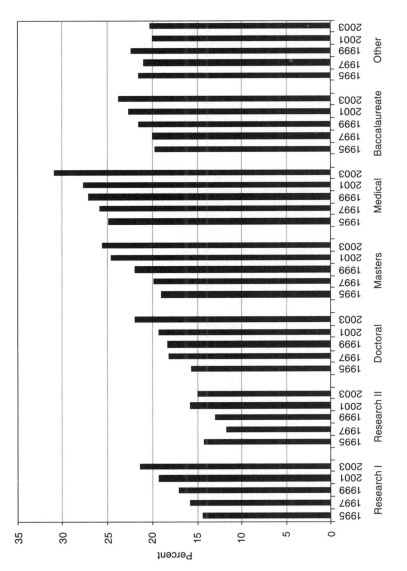

FIGURE A2-9 Percentage of faculty who were women by institution type, 1995-2003.
NOTE: Institutional classifications are distinct and do not overlap.
SOURCE: National Science Foundation, Survey of Doctorate Recipients, 1995-2003.

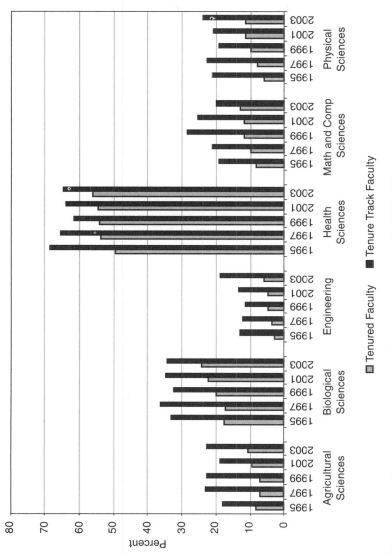

FIGURE 2-10 Percentage of faculty who were women by tenure status and field, 1995-2003.

SOURCE: National Science Foundation, Survey of Doctorate Recipients, 1995-2003. Tabulated by the NRC.

The highest proportions of women among the tenure-track faculty were found in health sciences.

Among both tenure-track and tenured faculty, women were proportionately more likely to be in medical colleges. About 15 percent of tenured faculty were women in Research I institutions; female tenured faculty were rarer at Research II institutions, but more prevalent at master's, doctoral, and baccalaureate institutions. The percentage of women among tenured faculty was growing at Research I institutions (see Figure A2-11).

Rank

Women were less likely to occupy senior positions in academia than men. Using the SDR, the committee examined rank by comparing the gender of faculty who were assistant, associate, and full professors, by academic discipline separately. Figure A2-12 shows the following results:

- Women comprised over 50 percent of all full professors in health sciences, 20 percent in biological sciences, and 10 percent or less in other fields, with engineering having the lowest proportion of female full professors.
- The percentage of women among full professors appeared to be rising or remaining level in each field.
- Women comprised almost 60 percent of all associate professors in health sciences, approximately 30 percent in biological sciences, and less than 20 percent in other fields, with engineering having the lowest proportion of female associate professors.
- The percentage of women among associate professors appeared to be rising or remaining level in many fields, but not in agricultural sciences and not in the health sciences.
- Women comprised 65 percent of all assistant professors in health sciences, 39 percent in biological sciences, between 25 to 27 percent in mathematics, computer, and physical sciences, with engineering having the lowest proportion of female assistant professors (less than 20 percent).
- The percentage of women among assistant professors appeared to be roughly steady in each field.

This analysis was then repeated, focusing on institution types. Figure A2-13 shows the following results:

- Women comprised about 25 percent of full professors at medical colleges and about 12 percent at Research I institutions.

263

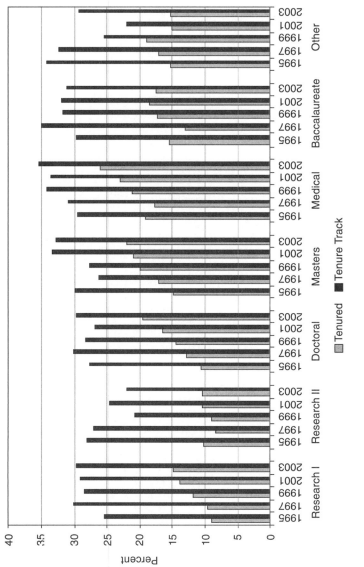

FIGURE A2-11 Percentage of faculty who were women by tenure status and institution type, 1995-2003.
SOURCE: National Science Foundation, Survey of Doctorate Recipients, 1995-2003. Tabulated by the NRC.

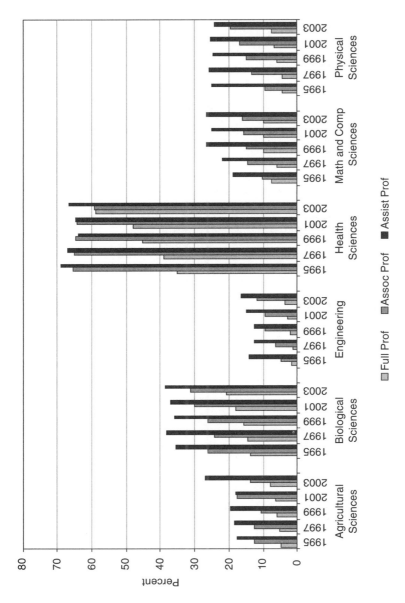

FIGURE A2-12 Percentage of faculty who were women by rank and field, 1995-2003.
SOURCE: National Science Foundation, Survey of Doctorate Recipients, 1995-2003. Tabulated by the NRC.

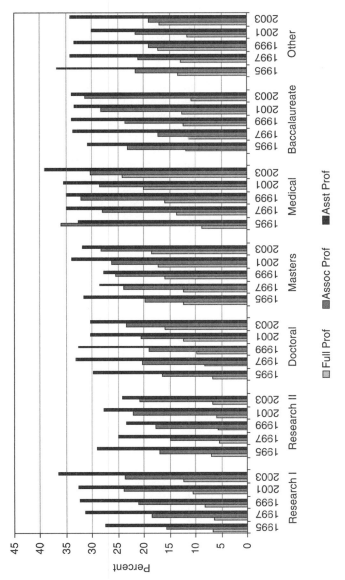

FIGURE A2-13 Percentage of faculty who were women by rank and institution type, 1995-2003.
SOURCE: National Science Foundation, Survey of Doctorate Recipients, 1995-2003. Tabulated by the NRC.

- The percentage of women among full professors appeared to be rising or remaining level in each field.
- Women comprised approximately 30 percent of associate professors at medical colleges and at baccalaureate institutions and approximately 24 percent at Research I institutions.
- The percentage of women among associate professors appeared to be rising or remaining level at each type of institution, except at medical colleges, where the trend was less clear.
- Women comprised between 35 and 40 percent of assistant professors at medical colleges. For assistant professors, there were more similarities across institution type. At each institution type, the proportion of women among assistant professors tended to be around 30 to 35 percent, excepting Research II institutions, which were lower.
- The percentage of women among assistant professors was rising at Research I institutions and at medical colleges but was less clear at other types of institutions.

Appendix 2-2
Previous Research on Factors Contributing to
Gender Differences Among Faculty

This appendix describes research on women in academic science engineering that provided a framework for the development of the 2004 and 2005 faculty and departmental surveys. Drawing primarily on studies conducted by individual institutions and the analyses of individual researchers, the research results suggested several possible reasons why women continued to represent a small segment of faculty—reasons that provided suggestions for survey questions and data needed to assess possible disparities.

TYPES OF RESEARCH

A survey of the literature uncovered many books and articles examining gender in academia, most of which examined gender issues either at the institutional or the individual level. Institutional factors focused on structures, processes, and policies, or the way institutions, departments, and faculty collectively functioned. Individual factors focused on characteristics of faculty members themselves. Many studies focused on one side or the other; fewer attempted to take both elements into account.

Institutional Studies

Individual universities and colleges have often conducted institutional research on salary, climate, or gender equity. One of the more well-known, but certainly not the first, gender equity studies was conducted by the Massachusetts Institute of Technology (MIT) in 1999. In recent years, more and more schools have conducted standalone gender equity reports.[1] Such reports typically collect and analyze data from institutional sources, including number of faculty in various departments or schools, disaggregated by gender. Several studies have collected new data by conducting on-campus surveys or focus group meetings on topics such as work/life policies, salary equity, climate, or faculty satisfaction. Interview-based approaches allow for questions to be raised on a wide variety of issues, including perceived treatment of self and colleagues, job satisfaction, and characterization of work activities.

---

[1] Reports for 80 of the 88 Research I institutions were collected and posted to the National Academies' Committee on Women in Science and Engineering (CWSEM) homepage, located at http://www7.nationalacademies.org/cwse/1gender_faculty_links.html.

Studies by Individual Researchers

Many scholars and researchers have carried out studies either using some of the national data sets or by collecting new information from surveys of faculty. As of 2004, there was a rich body of literature comparing various outcomes in the academic workforce by gender, focusing on a variety of factors:

- **Salary** (e.g., Barbezat, 2002; Becker and Toutkoushian, 2003; Ginther, 2001; and Perna, 2003c),
- **Supplemental earnings** (e.g., Perna, 2002),
- **Job satisfaction** (e.g., August and Waltman, 2004; Hagedorn et al., 1999; Olsen et al., 1995),
- **Productivity** (e.g., Porter and Umbach, 2001; Sax et al., 2002),
- **The probability of being in a tenure-track position** (e.g., NRC, 2001a; NSF, 2004d; Olson, 2002)
- **The probability of having tenure** (e.g., Ahern and Scott, 1981; Benedict and Wilder, 1999; NRC, 2001a; Perna, 2001a),
- **The probability of being an assistant or associate or full professor** (e.g., NRC, 2001a; NSF, 2004d; Olson, 2002; Ransom and Megdal, 1993),
- **The probability of being granted tenure** (e.g., Kahn, 1993),
- **The probability of being granted a promotion** (e.g., Ahern and Scott, 1981; Ginther, 2001),
- **Time to promotion** (e.g., Ginther, 2001),
- **Work activities, that is, time spent on research, teaching, and service** (e.g., Ahern and Scott, 1981),
- **Perceptions of (in)equality** (e.g., Robst et al., 1998), and
- **The likelihood of being retained or of leaving a faculty position** (e.g., Rosser, 2004; Zhou and Volkwein, 2004).

A 2003 literature review conducted by the National Science Foundation noted 15 studies on gender differences in rank and tenure and identified 13 studies focusing on gender differences in earnings in nationwide samples as well as several more studies employing a single-institution sample. Barbezat's (2002) "History of Pay Equity Studies" is another noteworthy review, which surveyed a number of studies on pay issues. A number of scholars used the Survey of Doctorate Recipients (SDR) to study gender differences (Ahern and Scott, 1981; Farber, 1977; Ginther, 2001; Kulis et al., 2002; NRC, 2001a; Olson, 2002) while other scholars employed data from the National Survey of Postsecondary Faculty (Bradburn et al., 2002; Glover and Parsad, 2002; Nettles et al., 2000; Perna, 2001c; and Toutkoushian, 1998a and b).

Examples of studies relying on original data collection include a study undertaken by Nelson and Rogers (2005), which looked at the number of male and female faculty members, by rank, at "top 50" departments in several fields. Several scholars turned to their own or a selection of institutions and collected data from institutional research offices, focus groups, or surveys to study this issue (e.g., Montelone et al., 2003; Nerad and Cerny, 1999a; Rosser, 2004; Trower and Bleak, 2004).

Limitations of Cross-Sectional Data Sources

Four major limitations to these types of cross-sectional data sources should be noted. First, although the academic career pathway is a longitudinal process, much of the data available cannot follow the same individual over a long period of time. Some faculty are surveyed in more than one SDR, but the SDR is not a panel study, even though it is longitudinal in its tracking of cohorts. For university studies, it is also possible that faculty would be in more than one survey. Longitudinal data that cover most of an individual faculty member's career are rare; the most consistently available data are snapshots of faculty at different points in their careers, taken at the time of the survey.[2]

Large gaps exist between the time periods selected for data collection. While some data collection occurs annually, such as salary surveys conducted by the American Association of University Professors (AAUP) or the American Chemical Society's (ACS's) survey of top 50 chemistry departments, most of the data available are not collected annually. Many university gender equity studies appear to be one-time events. The SDR is biennial.[3] The NSOPF has been conducted every 5 years since 1988, most recently in 2004.[4]

Second, the data may be biased or certain data points omitted. Doctoral graduates, for example, who fail to be hired and faculty who leave a university before or after tenure or promotion are less likely to be surveyed. The faculty who leave may exhibit different characteristics than the faculty who stay. As a result, analysis is likely to be restricted to the population of faculty who may be termed "successful" but does not represent all faculty. And it does not allow us to address other critical factors playing a significant role in determining the career paths of men and women in academia. Also, as these survey results are self-reported, data on productivity and job satisfaction may be biased, or faculty may simply misremember specific quantitative information from earlier stages of their career.

Third, comparability across studies is a major limiting factor, both in comparing surveys from the same series undertaken in different years and comparing

---

[2] This is part of the reason why most of the statistical analyses carried out use regression. A few scholars have used event history or hazard models. See for example Weiss and Lillard (1982), Kahn (1993), and Ginther (2001). See Allison (1984) for an introductory description of the methodology.

[3] Conducted on odd numbered years until 2003, thereafter on even numbered years, beginning in 2006.

[4] The National Center for Education Statistics also conducted a survey of department chairs during the 1988 NSOPF, but the chairs survey was only done this one time.

different surveys. In the case of the SDR and NSOPF, both of which have been carried out multiple times, the questions, how the survey is implemented, the sample size, and the response rate may all change. The NSF notes regarding the SDR:

> There have been a number of changes in the definition of the population surveyed over time. For example, prior to 1991, the survey included some individuals who had received doctoral degrees in fields outside of S&E or had received their degrees from non-U.S. universities. Since coverage of these individuals had declined over time, the decision was made to delete them from the 1991 survey. The survey improvements made in 1993 are sufficiently great that SRS staff believe that trend analyses between the data from the 1990s surveys and the surveys in prior years must be performed very cautiously, if at all.[5]

A more difficult task is comparing several university studies. Myriad approaches have been taken by universities in evaluating and assessing characteristics of their faculty, but concerns over comparability somewhat reduce the usefulness of the information gathered.

Fourth, in the interest of preserving confidentiality, surveys often provide aggregated information rather than the raw (i.e., individual) data. Certainly confidentiality is critical, but it means that some studies are less transparent in describing how the study was conducted and who was surveyed, making it more difficult to replicate or disaggregate the data and examine it differently. Readers are constrained by the findings reported by the scholars who put the data together.

SELECTED FACTORS CONTRIBUTING TO
GENDER DIFFERENCES AMONG FACULTY

Numerous factors have been used in the past to assess the status of male and female faculty in their careers. Characteristics that are often explored, aside from gender, are age, marital and family status, citizenship, field of study, educational experience (including highest degree and doctorate-granting institution), and employment experience (including number and types of previous jobs and characteristics of a faculty member's current position, such as rank or tenure status). The research on a few of these factors is highlighted here.

Relative Age of Women Faculty

In general, women as a group were younger than male faculty. Women are more recent entrants into academia than men, therefore women's representation among academic faculty was conditioned not only on the number of new Ph.D.s being granted to women, but also on the initial age and sex composition of faculty members and changes in the number of faculty positions (Hargens and Long,

---

[5] "Survey Methodology: Survey of Doctorate Recipients," NSF Web site at http://www.nsf.gov/sbe/srs/ssdr/sdrmeth.htm [accessed on March 17, 2004].

2002). Moreover, "while new cohorts of Ph.D.s entering the academic market-place are increasingly female, each new cohort is only a small proportion of those currently employed. Consequently, the move toward parity in the representation of women must occur slowly" (NRC, 2001:132). Hopkins (2006:16) gave an example in the case of MIT:

> In part, the small number of women faculty in [the Schools of] Science and Engineering can be explained by (1) the fact that the "pipeline" began to fill only about 40 years ago; and (2) faculty turnover rates are slow, with many faculty who achieve tenure staying at MIT for 30-40 years. Only about 5% of the MIT faculty leaves each year due to retirement, failure to achieve tenure, or other factors. At this rate, and assuming a 50% tenure rate, it would take approximately 40 years for a department that had no women faculty to have a faculty that has the same percentage of women as the Ph.D. pool.

As the NSF (1999:99) notes: "many of the differences in employment characteristics between men and women are partially due to differences in age. Women in the science and engineering workforce are younger, on average, than men: 18 percent of women and 12 percent of men employed as scientists and engineers were younger than age 30 in 1995." Since women faculty are younger, they have had, on average, less opportunity to receive tenure or a promotion, making career age a vitally important factor to control for in assessments of gender disparities in rank and tenure status (see e.g., NRC, 2001a; Olson, 2002).

Family Issues

Marital status and the presence of children were often mentioned as critical to assessing gender differences.[6] Rosser (2003) surveyed women who received an NSF POWRE award between 1997 and 2000. She found that "overwhelming numbers of survey respondents found 'balancing work with family' to be the most significant challenge facing women scientists and engineers. Interestingly, the responses remained remarkably similar across disciplines: balancing work with family responsibilities was the major issue for women from all the fields of study covered by the survey." Spouses and children presented competing demands for time on the part of a faculty member and might bring additional actors or considerations into decision making. These competing demands may have meant that some faculty had less human capital, experience, or productivity; or that applicants for academic positions were more constrained in where they applied because of family or the spouse's employment considerations (often referred to as geographic mobility or the two-body problem).

Did these factors affect men and women similarly? Research suggests that the answer was no. Women were more likely to be negatively affected by mar-

---

[6] Interestingly, research is adding care of older family members—for similar reasons as care of children (e.g., Sax et al., 2002).

riage and the presence of children. The NRC (2001a) found some evidence that being married with young children helped men but hurt women in terms of their academic career. The size of this effect had been shown to increase for men and to decrease for women. Xie and Shauman (2003) and Mason and Goulden (2002) found that marriage and family also negatively affected women pursuing science and engineering careers.

Toutkoushian (1998a:515) laid out an hypotheses as to why the effect of marital status on faculty salary might differ by gender: on the supply side, since women "often bear the majority of child-rearing responsibilities in American society, married women may be more likely than married men to interrupt or reduce their time allocation to their career," or "married women may accept lower wages in order to find employment at the same institution as their spouses." On the demand side, "institutions may make higher salary offers to married men than to married women on the premise that married men are typically the breadwinners of the family and thus have a greater need for higher salaries." Using NSOPF:93 to analyze faculty salaries, Toutkoushian found that the return on marriage for men was statistically significant and positive, but there was no corresponding return for women.

Sax et al. (2002:426) focused on the role of family-related variables in research productivity. Specifically, they asked: "Do marriage, children, aging parents, and other family-related factors influence faculty research productivity?" and "Is the nature of family-related factors dependent on gender or tenure status?" They analyzed data from the 1998-1999 Higher Education Research Institute Faculty Survey. They found, first, that male faculty were more productive than women, when compared at increasing levels of output over 2 years, i.e., a greater percentage of women than men produced zero publications, while a greater percentage of men than women produced five or more publications. However, Sax et al. found that "family variables contributed little or nothing to the prediction of faculty research productivity. More important were professional variables such as academic rank, salary, orientation toward research, and desire for recognition" (p. 435). Sax et al., hypothesized the lack of effect may have resulted because women who had children were able to do more with their limited time and reduce their time in activities outside of work and home (i.e., leisure time).

Perna (2003a:2) used the NSOPF:99 "to examine the ways in which parental status, marital status, and the employment status of the spouse are related to two outcomes, tenure and promotion, among college and university faculty." In an earlier study drawing on data from the NSOPF:93, Perna (2001c, cited in 2003a:3) "found that parental and marital status were related to employment status among junior faculty and that the relationships were different for women than for men. Men appeared to benefit from having children, as men with at least one child were less likely to hold a full-time, non-tenure track position than they were to hold a full-time, tenure track position." In this study, Perna found measures of family ties are related to tenure status and academic rank, but the contribution of family ties to tenure status and academic rank was different for women than for men.

Contrary to expectations based on economic and social capital perspectives, having dependents and having a spouse or partner employed at the same institution were both unrelated to tenure and rank among women faculty at 4-year institutions in the fall of 1999. In contrast, men appeared to have benefitted in terms of their tenure status and academic rank from having dependents and in terms of their academic rank from being married. Compared with other men, men without dependents were substantially less likely to hold tenured positions and were more likely to hold the lowest academic ranks of instructor, lecturer, and 'other.' Men also appeared to benefit in terms of their academic rank from being married. Specifically, men with a spouse or partner who was employed at the same institution were less likely to hold the lowest ranks of assistant professor and instructor, lecturer, or other rank than they were to hold the highest ranks of full and associate professor. Men with a spouse or partner who was not employed in higher education were more likely than other men to hold the rank of full professor.

Kulis and Sicotte (2002:2) examined "whether women are disproportionately drawn to large cities, areas with many local colleges, and the regional centers of doctoral production." Reviewing the literature, they suggested, regardless of academic achievement, wives in dual-career households were more likely to be the "trailing spouse" or "tied migrant" whose career suffered after a move, or were the one who was constrained from moving to a more advantageous career destination (p. 6). To test such hypotheses, they turned to the 1998 SDR. Their findings were essentially that women were congregated in fewer geographical areas. Women "scientists overall have more geographically constrained careers in academia, even controlling for marital status, parental responsibilities, and age" (p. 21). Women in these areas also had reduced career outcomes compared with men."

Mason and Goulden (2002) conducted research on "family formation and its effects on the career lives of both women and men academics from the time they receive their doctorates until 20 years later." They employed data from the SDR for 1973-1999. They found "in the sciences and engineering, among those working in academia, men who have early babies are strikingly more successful in earning tenure than women who have early babies. Surprisingly, having early babies seems to help men; men who have early babies achieve tenure at slightly higher rates than people who do not have early babies. Women with early babies often do not get as far as ladder-rank jobs." Data from the analysis of the SDR suggested many married women with children indicated that they were considering leaving academia.

Institutional Policies and Practices

Previous research on the role of institutions in gender differences among their faculty consisted of two different approaches. One approach focused on structural differences among institutions, arguing that such variables as the type of institution, whether it was a public or private institution, its prestige, whether it was unionized, and even its geographic region could explain some of the differences between male and female faculty members. A more challenging approach focused on the way in which universities worked—hire faculty, grant tenure, and promote faculty—arguing that these policies and procedures could be biased against particular groups of people (see e.g., Gibbons, 1992b; Menges and Exum, 1983; Steinpreis et al., 1999; and Valian, 1998; 2004).[7]

An example of an important policy affecting women's academic careers was stopping the tenure clock. By 2004, many universities had such policies in place, but some studies found that faculty were hesitant to make use of such a policy. For many women, the fear that taking an extension might cause their senior colleagues to view them more negatively and hurt their career—an effect not conclusively documented—was sufficient to dissuade them from taking this option (Bhattacharjee, 2004b).

Policies about hiring spouses were also seen as relevant in both hiring and retention of women, as women were more likely than men to be married to other academics. Equally important were policies on child care and parental leave. According to researchers, creating spousal hiring programs and establishing parental leave policies and child care were practices that "would make academic institutions more attractive to prospective candidates of either gender" (Sullivan et al., 2004).

This review of previous literature and research reflects the opinions at the time of this study's surveys of faculty and departments. They should help to assess the climate at research-intensive institutions at that time and may be helpful in assessing how effective the efforts of these institutions have been since then to improve the representation of women S&E faculty.

---

[7] A review by the Women in Science & Engineering Leadership Institute (WISELI) at the University of Wisconsin-Madison titled, "Reviewing Applicants: Research on Bias and Assumptions" identified several studies suggesting that female candidates may have a tougher time. Available at http://wiseli. engr.wisc.edu/doc/BiasBrochure_2ndEd.pdf [accessed on October 7, 2008].

Appendix 3-1
Review of Literature and Research on
Factors Associated with a Higher Proportion of
Female Applicants

This appendix examines prior research on the factors hypothesized to be associated with the proportion of female applicants for faculty positions. The focus is on departmental or institutional characteristics since this study's survey data contain little information about the individual applicants, apart from their gender. A review of previous research included topics on departmental climate, work-life balance and family-friendly policies, geographic location, departmental prestige, and public versus private institutions. In addition, we examined the relationship between availability of women in the Ph.D. pool and the percentage of female applicants.

STATUS OF WOMEN FACULTY OVERALL IN 2003

A review of previous research at the time the surveys were conducted showed that the proportion of female science and engineering (S&E) faculty at Research I (RI) institutions was rising but had yet to reach parity in reference to the proportion of S&E doctorates awarded to women. From 1979 to 2003, the percentage of female S&E assistant professors at these institutions grew from 11 percent to over 35 percent; the percentage of female S&E associate professors rose from 5 percent to 24 percent; and the percentage of female S&E full professors rose from about 2 percent to about 11 percent.[1] These aggregate trends masked substantial variability across departments. Some disciplines, such as biology, had attracted many more female faculty than others, and within a specific discipline, some departments had attracted many female faculty while others still have no women among their faculty members (e.g., Ivie and Ray, 2005; Kuck et al, 2004; Nelson and Rogers, 2005). Additionally, there had been some concern that while earlier efforts beginning around 2000 had increased female representation, those efforts may had stalled out.

Both the overall rate of growth in the percentage of S&E faculty who were women and the variation among departments, disciplines, and institutions may be partly attributable to the hiring process.

---

[1] Data for 1979 are from NRC (2001a) and were calculated by taking total number of male and female faculty at Research I institutions and subtracting male and female faculty at Research I institutions who were in social and behavior sciences. Data for 2003 are also from the Survey of Doctorate Recipients (SDR) as calculated by staff, using the same definition of S&E.

AVAILABILITY OF WOMEN IN THE PH.D. POOL

The potential applicant pool consists of those individuals who could apply for one or more positions. In practice, universities know only the number of applicants who apply for any particular position for which they are recruiting, and the actual potential candidate pool remains unknown. Typically, the number of women receiving Ph.D.s in a field is used as a proxy for the eligible pool of women.[2]

As noted in Chapter 2 and Appendix 2-1, the number of women receiving Ph.D.s in S&E had grown significantly over the years—both numerically and as a proportion of all those receiving doctorates in S&E. On average, over the period from 1999 to 2003, the 5-year period preceding the survey's focus, Research I institutions awarded women 45 percent of the Ph.D.s in biology, 32 percent in chemistry, 18 percent in civil engineering, 12 percent in electrical engineering, 25 percent in mathematics, and 14 percent in physics. In 2003, 4,005 women received Ph.D.s from all doctorate-granting institutions for the six fields studied (see Appendixes 3-4 and 3-5):

- 2,598 Ph.D.s (45.7 percent) in biology;
- 647 Ph.D.s (31.8 percent) in chemistry;
- 125 Ph.D.s (18.7 percent) in civil engineering;
- 179 Ph.D.s (12.3 percent) in electrical engineering;
- 263 Ph.D.s (26.5 percent) in mathematics and statistics; and
- 193 Ph.D.s (18.0 percent) in physics.

A majority of doctoral degrees are awarded by the 89 Research I institutions (see Appendix 3-6).

On average, one might expect disciplines with higher proportions of female doctorates would also see higher proportions of female applicants. Thus, a reasonable expectation is women will make up a larger proportion of applicants to positions in biology and chemistry, followed by mathematics, civil engineering, physics, and electrical engineering. This seems to be the case generally for tenure-track jobs in our study (with the exception that the rank order positions of chemistry and mathematics are reversed, but it does not hold at all for tenured jobs.

---

[2] This measure is deficient in two ways. First, the potential applicant pool includes postdocs, individuals with Ph.D.s from foreign institutions, individuals from outside academia, and individuals with current academic positions who are interested in switching to a new position (Ehrenberg, 1992). For example, in a study of physics hires in 2000, Kirby et al. (2001) found that 34 percent of new hires in doctorate-granting institutions had earned Ph.D.s outside of the United States. Likewise, in computer science (Zweben, 2005:10), for 2003-2004: "Thus, more than 75% of the faculty hires made this past year by Ph.D.-granting CS/CE [computer science/computer engineering] departments appear to have been new Ph.D.s, with the rest consisting of a combination of faculty who changed academic positions, persons joining academia from government and industry, new Ph.D.s from outside of North America and from disciplines outside of CS/CE, and non-PhD. holders (e.g., taking a teaching faculty appointment)." Second, it fails to account for the preferences of doctorates.

A commonly heard gender-based explanation offered to account for differences between the proportion of women in the Ph.D. pool and the proportion among applicants for Research I positions is that many women S&E doctorates may not be interested in academic positions at Research I institutions. It is the case, as noted in Chapter 2, that many women Ph.D.s were employed outside academia, and within academia, many women were employed at institutions other than Research I institutions. This was not unexpected since the 89 Research I institutions make up only a small part of higher education institutions.

Fox and Stephan (2001) examined the preferences of 3,800 doctoral students in chemistry, computer science, electrical engineering, microbiology, and physics. Overall, 36 percent of students had a preference for academic research, compared with 19 percent, who indicated a preference for academic teaching. In every case, the proportion of women preferring academic teaching was greater than that of men. Men strongly preferred academic research in chemistry, microbiology, and computer science, more than women did.

Sears (2003) conducted a survey of 1,105 graduate students from 24 math and science programs at the University of California at Davis, with a focus on comparing students' initial career goals when they began graduate school with their current career goals. A crucial finding was "more men than women began graduate school with plans to work in research universities (84% of men, 71% of women), and during graduate school, more women than men abandoned this goal" (p. 172). Additionally, men, more than women respondents, were attracted to research universities. Bleak et al. (2000), in a survey of recently hired faculty, found men were more likely to apply to research universities than women. Data collected by the American Chemical Society also suggested women were choosing 4-year institutions over research universities (Brennan, 1996).

Why might women be less interested in positions in research universities? In general, women graduates may perceive the climate to be less welcoming, perhaps based on their perceptions of how they were treated in graduate school and their perceptions of how female faculty were treated. There was evidence that female graduate students may perceive the social or cultural context of doctoral education in S&E differently than male graduate students do. In a survey of 3,300 students in chemistry, computer science, electrical engineering, and physics, conducted during 1993 to 1994, Fox (2001a) found:

- "Women are less likely than men to report that they are taken seriously by faculty and that they are respected by faculty" (p. 658).
- "In research groups, compared to men, women report that they are less comfortable speaking in group meetings" (p. 659).
- "Women report collaborating with fewer men graduate students and men faculty members in research and publications during the three preceding years" (p. 659).

- "Men are more apt to have received help [from advisers] in these areas [learning to design research, write grant proposals, coauthor publications, and organize people] across types of departments" (pp. 659-660).
- "Women are also more likely than men to report that they view their relationship with their adviser as one of 'student-and-faculty' compared with 'mentor-mentee' or 'colleagues,' which may suggest greater formality and social distance for women students" (p. 670).
- In terms of outcomes, men "publish more papers and are more likely to report that they will receive their degrees" (p. 660).

Fox (2001a:660) concluded "if women are constrained within the social networks of science—in departments or in the larger communities of science—this restricts their possibilities not simply to participate in a social circle but, more fundamentally, to do research, to publish, to be cited—to show the marks of status and performance in science (Fox 1991)." The level of socialization may affect the ability of individuals to find a position. In addition, the degree of integration into a department's life as well as closeness with a faculty member may impact whether one learns important details about available academic positions or feels encouraged to apply.

DEPARTMENTAL CLIMATE

One of the reasons women might not apply to RI institutions is there is a perception that these schools have a reputation for not being female-friendly. Female students may experience a chilly climate in graduate school or may perceive that some female faculty find obstacles when pursuing their careers and, as a consequence, may opt to embark on a career elsewhere (Brennan, 1996). The committee considered a number of variables potentially reflecting the department climate for women.

REPRESENTATION OF WOMEN IN DEPARTMENTS

The committee hypothesized having a larger proportion of women in a department might be taken by female potential applicants as a signal of a "woman-friendly" department. Thus, the percentage of women applicants would be expected to be positively correlated with the percentage of women already on the faculty. However, prior research indicated this relationship may not have been linear. In their study of 93 academic positions, Yoder et al. (1989) found "departments with more than half women did not appear to be very willing to hire additional women, while departments with moderate numbers of women were. Ironically, departments with few or no women were almost as unlikely as departments numerically dominated by women to hire a woman" (p. 272). Yoder et al. explained this outcome by noting, in departments with few female faculty, women had little power to influence group decision making, a version of the critical mass hypothesis. In departments with some women—between 16 and 35 percent, women could form alliances and coalitions to influence the group. When women achieved balance in a department, job hires became less about equity, and men and women were hired at equal rates.

REPRESENTATION OF WOMEN ON THE SEARCH COMMITTEE

Female applicants may also take the presence of women on the search committee as a sign of a more female-friendly environment. At meetings of professional societies and as a focal point of hiring efforts, the search committees may be very visible, and having a female search committee chair may lead to greater efforts to recruit female applicants.

BALANCING WORK-LIFE AND FAMILY-FRIENDLY POLICIES

It may be more difficult to balance family and career at a Research I institution (Sears, 2003), which may discourage women from applying for RI positions. Marital status and the presence of children are often mentioned as critical to assessing gender differences.

Institutions with spousal support policies and child-care and family leave policies might be more attractive to female doctorate recipients. For example, readily available child care may make a greater positive difference in the lives of female faculty than male faculty. Leave policies are another institutional policy that may affect female and male faculty differently. Two types of leave include maternity leave, which is a standard benefit at universities, and longer, parental leave (Yoest, 2004). Some universities also have workload relief policies (typically a reduction in teaching and service responsibilities) for new parents. Spousal policies can take on a number of different forms. Wolf-Wendel et al. (2003:163) suggested six broad approaches to "help spouses and partners of academics find suitable employment." These were relocation assistance, hiring a spouse or partner into an administrative position, hiring a spouse or partner into a non-tenure-track position, creating a shared position, creating a joint position with a nearby institution, or creating a tenure-track position for the spouse or partner. Again, spousal policies were most relevant to hiring issues. The availability of these policies may affect the probability that women will apply for particular positions.

PUBLIC VERSUS PRIVATE UNIVERSITIES

Public universities are often thought to do more to foster diversity than private institutions. This is because these institutions have more state oversight and may be more transparent. Insofar as this is widely believed, women may be more likely to apply for positions at public than at private research universities.

GEOGRAPHIC MOBILITY

Marital and family status present competing demands for time on the part of a faculty member and may bring additional actors or considerations into decision making. Female applicants for academic positions may be more constrained in where they can apply. Taking into consideration children and their education and a spouse's employment preferences and opportunities all mean women may be more likely to take other interests into account, aside from their own preferences.

A special subset of the family-work problem concerns dual-career couples, also known as the two-body problem (Wolf-Wendel et al., 2003). "Nearly 38% of women chemists are married to a chemist or other scientist, according to the 1995 ACS survey. . . . Just shy of 21% of male chemists are married to a scientist" (Slade, 1999:61). "According to figures from the American Institute of Physics, 44% of married women in physics are married to other physicists—and another 25% to some of scientist. A remarkable 80% of female mathematicians are married to other scientists or engineers, along with a third of female chemists" (Gibbons, 1992c). It may be difficult to find two academic openings at the same department. Additionally, trying to find two jobs at a Research I institution is often perceived as more difficult than at other types of institutions.

The question here is: Are women as mobile as men or are there factors constraining where a woman can work? If so, then men may be able to apply to more jobs than women, who may be clustered in applicant pools for a smaller number of jobs. Research supports this view. The general geographic mobility argument is that changing jobs for many academics is a positive (upward mobility), and the academic labor market is national so academics should be flexible to take advantage when opportunity knocks. Women are less able to do this, largely because of marriage. The careers of married women are likely to take a backseat to the careers of their husbands (Marwell et al., 1979; Rosenfeld and Jones, 1987). Rosenfeld and Jones argue that single women might also be geographically constrained. They may prefer large cities, which offer more possibilities for various types of social networks.

As noted in the Appendix 2-1, Kulis and Sicotte (2002:2) suggested careers of women are more likely to be geographically constrained. Their analysis indicated female faculty are more likely than male faculty to reside in doctoral production centers, areas with large clusters of colleges, and large cities. They also found women in these areas had reduced career outcomes compared with men. "Geographic constraints appear to be more disadvantageous for women, and the career advantages associated with certain locations generally seem to help women much less than men. For example, compared to men living in the same areas and women living elsewhere, women located in high doctoral production regions are less likely to have tenure and more likely to work part time. Both men and women in large cities are more likely to be employed off the tenure track, but the women occupy these jobs far more often than the men" (p. 24). For our purposes, the relevant consequence of this argument was that women are more likely to consider geography when deciding where to apply for academic jobs.

Data from more recent surveys continued to note the differential importance of location for women. In its survey on chemists, the American Chemical Society (Ellis, 2001:23) reported, "in searching for work, the inability to relocate is cited much more often by women than by men as a constraint." Among those chemists who were unemployed and actively seeking positions, "close to 37 percent of women in 2000 noted that it was because of an inability to relocate, whereas only 27.4 percent of men listed the same reason. Just over 15 percent of women, and 9.1 percent of men, said it was because of family responsibilities. The percentage of women who reported that they placed no job restrictions in their job search was 28.3 percent as opposed to 48.8 percent of men (Kreeger, 2001:14). Bleak et al. (2000:14) noted recently hired female faculty placed more emphasis than male faculty on location of the institution and employment opportunities for their spouse or partner. Sears' (2003:175) study of graduate students in science and math programs at the University of California, Davis found "women were much more likely than men to report that location was an important factor in job selection because of the location of their spouses' jobs or their desire to be close to family and friends."[3]

An important consequence is that women may not choose to apply to as many jobs as men, even among the Research I institutions. Women, especially married women, could be less likely to apply to RI institutions in smaller towns, where there are fewer opportunities for spouses. A second important consequence of mobility constraints might be that search committees are less likely to offer women positions if the committee believes the woman will not accept the offer.

PRESTIGE

The most prestigious institutions tend to do least well in recruiting female faculty. "The higher up the academic-prestige ladder a university is, the fewer women it usually has in tenured faculty positions. Research released showed that while the nation is doing a good job of turning out women with research doctorates, the top 50 institutions in research spending are not doing such a good job of hiring them" (Wilson, 2004).[4]

The under representation of women in the most prestigious departments could result either from a lack of demand for female faculty in these departments or from a lack of supply of female candidates. Potential faculty may be likely to consider the reputation of both the department and the institution in deciding which jobs openings they will apply for. Some argue greater prestige may not always be seen as a positive attribute by female applicants. "Women just are not applying, "says

---

[3] Ellipses omitted.

[4] See also Bain and Cummings (2000).

Geraldine L. Richmond, who holds an endowed chair in chemistry at the University of Oregon. She argues "many top-notch science departments have 'toxic atmospheres' that suffocate women's enthusiasm for their work and steer them away from research careers. But women are also rejecting elite research universities for other reasons, like the fear that they will not have enough time for their families" (Wilson, 2004).

Kulis and Miller-Loessi (1992) offered a different rationale: higher prestige institutions seek to attract high-powered researchers. In the past, those would more likely be men. The authors noted women have been located outside informal prestige networks, making it harder for women to be recognized and recruited.

Steinpreis et al. (1999) simulated a hiring situation by sending 238 male and female academic psychologists one of four randomly selected versions of curriculum vitae (CV) along with a questionnaire about the qualifications of the candidate. The CV was drawn from a real-life, female scientist. Some versions of the CV contained a traditional male name; other versions, a traditional female name. The authors found "both male and female academicians were significantly more likely to hire a potential male colleague than an equally qualified potential female colleague. Furthermore, both male and female participants were more likely to positively evaluate the research, teaching, and service contributions of a male job applicant than a female job applicant with an identical record" (p. 522).

Several other studies reach the similar conclusion that female candidates may be at a disadvantage in both academic and nonacademic labor markets:

- Cole et al. (2004) randomly sent business school students' resumes to 40 employers, who were asked to rate the resumes on a number of criteria. They found male reviewers rated male applicants as having slightly more work experiences than female applicants (not statistically significant), while female reviewers rated male applicants as possessing significantly more work experience.
- Studies suggest women's professional work is discounted more so than for men. For example, a study of the outcomes of the peer-review system of the Swedish Medical Research Council for postdoctoral fellowships found the success rate for female applicants was less than half that of male applicants (Wenneras and Wold, 1997).

The situation applies not just to female versus male names as triggers, but also to female versus male appearance. In the music world, very few women were playing with top orchestras in the 1970s. Then orchestras changed how the audition occurred: the musician was hidden behind a screen and the stage was carpeted. The number of women successfully auditioning rose significantly (Koretz, 1997; Goldin and Rouse, 2000). Women seem to get rated harder than men do, both by men and women. However, one study did not find a disparity. In a review of editors, reviewers, and authors regarding manuscripts submitted to *JAMA* in 1991, the authors found that there were gender differences in how editors worked and how reviewers made recommendations, but they found final "manuscript acceptance rates did not differ across author gender and editor gender combinations" (Gilbert et al., 1992). Another study by Swim et al. (1989)—where the authors conducted a meta-analysis on studies drawing on the influential experiment conducted by Goldberg in 1968—demonstrated that women rated publications perceived to have been written by female authors less favorably than those thought to have been written by males.

This bias could occur because of at least two different kinds of stereotypes about women (Cole et al., 2004). Evaluators could have descriptive stereotypes. For example, they could believe women "don't have what it takes to succeed in competitive situations." Alternatively, evaluators could have prescriptive stereo-types. A woman perceived as behaving in an unfeminine way to get an academic position could be negatively evaluated for her behavior. In addition to broad gender stereotypes, gender stereotypes specific to the academic world, such as a perception that women are less mobile or less committed to the profession, may affect invitations to interview. Differences in the level of socialization among male and female graduate students and postdocs may further impact an aspiring faculty member by affecting the quality of letters of reference. This may be a significant problem. Trix and Psenka (2003), for example, found recommendation letters for women for medical faculty positions were shorter, less favorable, and focused more on women's teaching abilities than the letters for men.[5] In general, percep-tions regarding women, held by both men and women, may have a detrimental effect on hiring or career advancement (Valian, 1998).

---

[5] This is not a new problem. Stake et al. (1981) found letters of recommendation were more favor-able when the letter writers and the job seekers were of the same gender.

Appendix 3-2

Estimated Adjusted Mean Effects and Differences for the Probability That There Are No Female Applicants[a]

Differences Across Effect Levels	Estimated Mean Difference (Lower 95%, Upper 95% Confidence Limits)
Biology – Chemistry[b]	0.22 (–0.08, 0.51)
Biology – Mathematics	0.50 (0.01, 0.99)
Biology – Electrical engineering	0.23 (–0.12, 0.57)
Biology – Physics	0.22 (–0.11, 0.54)
Biology – Civil engineering	0.13 (–0.07, 0.34)
Tenured – Tenure-track	0.81 (0.71, 0.92)
Private institution – Public institution	0.66 (0.49, 0.84)
Top 10 department – Next 10 depts.	0.27 (0.10, 0.44)
Next 10 departments – Remaining depts.	0.81 (0.59, 1.03)
M – F search committee chair	0.24 (–0.16, 0.63)

[a] The sample size used to fit this model was 667. The effects fit were: (1) indicator variables for discipline (Biology, Chemistry, Civil Engineering, Electrical Engineering, Mathematics, and Physics, (2) indicator variables for Tenured, Tenure-track, (3) indicator variables for private institution, public institution, (4) indicator variables for top ten departments, second ten departments, and remainder, and (5) an indicator variable as to whether the committee chair was female.

[b] The estimated adjusted mean differences can be interpreted using Biology – Chemistry as an example. For those individuals in Biology, there is an estimated probability of having no female applicants given, or conditional on, the values for the remaining predictors in the logistic regression model. There is an analogous set of estimated conditional probabilities for Chemistry, again conditional on the predictors in the model. For each set of predictors, one can compute the difference of the estimated probabilities, and then one can average these differences in estimated probabilities over the estimated distribution of the predictors. The result is an estimated average difference of probabilities.

SOURCE: Departmental survey conducted by the Committee on Gender Differences in Careers of Science, Engineering, and Mathematics Faculty.

Appendix 3-3

Estimated Adjusted Mean Effects and Differences Based on the Modeled
Probability of the Percentage of Applicants That Are Female [a,b,c]

Effects	Mean (Lower, Upper 95% Confidence Limits)
Disciplines	
Biology	0.24 (0.20, 0.28)
Chemistry	0.17 (0.14, 0.21)
Mathematics	0.20 (0.16, 0.24)
Electrical engineering	0.10 (0.07, 0.13)
Physics	0.10 (0.09, 0.12)
Civil engineering	0.13 (0.10, 0.17)
Type of position	
Tenured	0.15 (0.12, 0.17)
Tenure track	0.15 (0.13, 0.17)
Tenured – tenure-track	1.00 (0.88, 1.12)
Type of institution	
Private	0.15 (0.13, 0.17)
Public	0.15 (0.12, 0.17)
Private – public	1.03 (0.87, 1.19)
Prestige of institution	
Top 10 institutions	0.16 (0.14, 0.18)
Next 10 institutions	0.14 (0 10, 0.19)
Remaining institutions	0.15 (0.13, 0.17)
Top 10 – second 10	1.12 (0.75, 1.48)
Top 10 – remaining	1.08 (0.94, 1.22)
Second 10 – remaining	0.97 (0.69, 1.25)
Gender of search committee chair	
Female chair	0.16 (0.13, 0.19)
Male chair	0.14 (0.12, 0.16)
Female – male chair	1.17 (1.01, 1.32)

Effects	Ratios of Means (Lower, Upper 95% Confidence Limits)
Differences across disciplines	
Biology – Chemistry	1.36 (1.10, 1.62)
Biology – Mathematics	1.21 (1.05, 1.37)
Biology – Electrical engineering	2.44 (1.61, 3.27)
Biology – Physics	2.30 (1.91, 2.69)
Biology – Civil engineering	1.80 (1.29, 2.32)
Chemistry – Mathematics	0.89 (0.69, 1.08)
Chemistry – Electrical engineering	1.79 (1.18, 2.39)
Chemistry – Physics	1.69 (1.39, 1.98)
Chemistry – Civil engineering	1.32 (0.92, 1.73)
Mathematics – Electrical engineering	2.02 (1.35, 2.68)
Mathematics – Physics	1.90 (1.57, 2.24)
Mathematics – Civil engineering	1.49 (1.06, 1.93)
Electrical Engineering – Physics	0.94 (0.64, 1.25)
Electrical – Civil engineering	0.74 (0.47, 1.01)
Physics – Civil engineering	0.78 (0.58, 0.99)
Type of position	
Tenured – tenure-track	1.00 (0.88, 1.12)
Type of institution	
Private – public	1.03 (0.87, 1.19)
Prestige of institution	
Top 10 – second 10	1.12 (0.75, 1.48)
Top 10 – remaining	1.08 (0.94, 1.22)
Second 10 – remaining	0.97 (0.69, 1.25)
Gender of search committee chair	
Female – male chair	1.17 (1.01, 1.32)

[a] The sample size used to fit this model was 667.

[b] The same effects were fit as in the table in Appendix 3-2.

[c] See note b in the table in Appendix 3-2.

SOURCE: Departmental survey conducted by the Committee on Gender Differences in Careers of Science, Engineering, and Mathematics Faculty.

Appendix 3-4

Estimated Adjusted Mean Effects and Differences Based on the Modeled
Probability of at Least One Female Candidate Interviewed[a]

Effects	Mean Odds Ratios (Lower, Upper 95% Confidence Limits)
Disciplines	
Biology	0.51 (0.25, 0.76)
Chemistry	0.80 (0.68, 0.91)
Mathematics	0.80 (0.64, 0.96)
Electrical engineering	0.84 (0.72, 0.96)
Physics	0.84 (0.73, 0.95)
Civil engineering	0.81 (0.66, 0.96)
Type of position	
Tenured	0.73 (0.59, 0.86)
Tenure-track	0.82 (0.74, 0.91)
Tenured – tenure-track	0.57 (0.22, 0.93)
Type of institution	
Private	0.79 (0.67, 0.92)
Public	0.77 (0.65, 0.92)
Private – public	1.17 (0.19, 2.16)
Prestige of institution	
Top 10	0.82 (0.72, 0.91)
Next 10	0.75 (0.59, 0.91)
Remaining institutions	0.77 (0.62, 0.92)
Top 10 – Next 10	1.50 (0.39, 2.62)
Top 10 – Remaining	1.33 (0.02, 2.65)
Next 10 – Remaining	0.89 (–0.13, 1.90)
Search committee chair	
Male	0.77 (0.62, 0.93)
Female	0.78 (0.71, 0.86)
Male – female	0.95 (0.16, 1.74)

Effects	Ratio Of Mean Odds Ratios (Lower, Upper 95% Confidence Limits)
Differences across disciplines	
Biology – Chemistry	0.26 (–0.04, 0.55)
Biology – Mathematics	0.26 (–0.08, 0.59)
Biology – Electrical engineering	0.19 (–0.06, 0.45)
Biology – Physics	0.19 (–0.04, 0.42)
Biology – Civil engineering	0.24 (–0.08, 0.56)
Chemistry – Mathematics	0.99 (–0.01, 2.00)
Chemistry – Electrical engineering	0.75 (0.06, 1.44)
Chemistry – Physics	0.75 (–0.01, 1.50)
Chemistry – Civil engineering	0.94 (0.03, 1.85)
Mathematics – Electrical engineering	0.76 (–0.11, 1.62)
Mathematics – Physics	0.75 (–0.03, 1.53)
Mathematics – Civil engineering	0.95 (–0.30, 2.19)
Electrical engineering – Physics	0.99 (0.07, 1.92)
Electrical engineering – Civil engineering	1.25 (0.13, 2.37)
Physics – Civil engineering	1.26 (0.10, 2.41)
Type of position	
Tenured – tenure-track	0.57 (0.22, 0.93)
Type of institution	
Private – public	1.17 (0.19, 2.16)
Prestige of institution	
Top 10 – Next 10	1.50 (0.39, 2.62)
Top 10 – Remaining	1.33 (0.02, 2.65)
Next 10 – Remaining	0.89 (–0.13, 1.90)
Search committee chair	
Male – female	0.95 (0.16, 1.74)

[a] The sample size used to fit this model was 667. For differences across effect level, the mean represents the ratio of odds ratios between the two factor levels.

[b] The same effects were fit as in Appendix 3-2.

SOURCE: Departmental survey conducted by the Committee on Gender Differences in Careers of Science, Engineering, and Mathematics Faculty.

Appendix 3-5

Doctoral Degrees Awarded by All Doctoral-Granting Institutions, by Field, Gender, and Year

Field	Gender	1999	2000	2001	2002	2003
Civil engineering	Female	89	88	111	120	125
Civil engineering	Male	495	466	482	504	544
Civil engineering	Percent Female	15.2%	15.9%	18.7%	19.2%	18.7%
Electrical engineering	Female	155	195	203	163	179
Electrical engineering	Male	1,310	1,339	1,372	1,223	1,276
Electrical engineering	Percent Female	10.6%	12.7%	12.9%	11.8%	12.3%
Chemistry	Female	632	624	628	647	647
Chemistry	Male	1,493	1,361	1,349	1,275	1,385
Chemistry	Percent Female	29.7%	31.4%	31.8%	33.7%	31.8%
Physics	Female	160	163	160	177	193
Physics	Male	1,103	1,040	1,036	946	882
Physics	Percent Female	12.7%	13.5%	13.4%	15.8%	18.0%
Mathematics and Statistics	Female	277	259	276	265	263
Mathematics and Statistics	Male	803	790	731	650	729
Mathematics and Statistics	Percent Female	25.6%	24.7%	27.4%	29.0%	26.5%
Biological sciences	Female	2,394	2,622	2,549	2,544	2,598
Biological sciences	Male	3,171	3,226	3,133	3,142	3,083
Biological sciences	Percent Female	43.0%	44.8%	44.9%	44.7%	45.7%

SOURCE: NSF, WebCASPAR.

Appendix 3-6

Doctoral Degrees Awarded by Discipline and Gender for Research I Institutions, 1999-2003

Field	1999		2000		2001		2002		2003	
	Female	Male	Female	Male	Female	Male	Female	Male	Female	Male
Biology	1,725	2,239	1,875	2,279	1,824	2,212	1,776	2,181	1,892	2,198
Chemistry	457	1,094	470	985	478	984	488	958	481	1,010
Civil engineering	78	386	72	384	95	365	86	400	89	422
Electrical engineering	124	1,015	143	1,045	154	1,076	133	939	138	981
Mathematics	194	625	189	613	202	558	196	507	184	589
Physics	108	862	122	759	113	792	122	739	138	693

SOURCE: NSF, WebCASPAR.

Appendix 3-7
Marginal Mean and Variance of
Transformed Response Variables

Data collected in the departmental and faculty surveys were used to answer various research questions in this report. Statistical analyses consisted essentially of fitting various types of regression models, including multiple linear regression, logistic regression, and Poisson regression models depending on the distributional assumptions that were appropriate for each response variable of interest. In some cases, the response variable was transformed so that the assumption of normality for the response in the transformed scale was plausible. Marginal or least-squares means were calculated (sometimes in the transformed scale) for effects of interest in the models.

TRANSFORMATIONS

We let y denote a response variable such as the proportion of women in the applicant pool or annual salary or number of manuscripts published in a year, and use x to denote a vector of covariates that might include type of institution, discipline, proportion of women on the search committee, etc. If y can be assumed to be normally distributed with some mean μ and some variance σ^2 then we typically fit a linear regression model to y that establishes that $\mu = x\beta$, where β is a vector of unknown regression coefficients.

When the response y is not normally distributed (for example, because y can only take on values 0 and 1) then we can define $\eta = x\beta$ and then choose a transformation g of μ such that

$$g(\mu) = \eta = x\beta.$$

For example, if the response variable is a proportion, the logit transformation

$$g(\mu) = \log\left(\frac{\mu}{1-\mu}\right)$$

is appropriate. When y is a count variable (as in the number of manuscripts published in a year) the usual transformation is the log transformation.

One approach to obtaining estimates of β is the method of maximum likelihood. Let $\hat{\beta}$ denote the maximum likelihood estimate (MLE) of β. A nice property of MLEs is invariance; in general, the MLE of a function $h(\beta)$ is equal to the function of the MLE of β, thus

$$\hat{h}(\beta) = h(\hat{\beta}).$$

In particular, if $\hat{\eta} = x\hat{\beta},$ then

$$\hat{\mu} = g^{-1}(\hat{\eta}).$$

The difficulty arises when we wish to also estimate the variance of $\hat{\mu}$ for example to then obtain a confidence interval around the point estimate $\hat{\mu}$. To do so, we typically need to resort to linearization techniques that allow us to compute an approximation to the variance of a non-linear function of the parameters. A method that can be used for this purpose is called the Delta method and is described below.

LEAST-SQUARES MEANS

Least-squares means of the response, also known as adjusted means or marginal means can be computed for each *classification* or *qualitative* effect in the model. Examples of qualitative effects in our models include type of institution (two levels: public or private) discipline (with six categories in our study), gender of chair of search committee, and others. Least-squares means are predicted population margins or within-effect level means adjusted for the other effects in the model. If the design is balanced, the least-squares means *(LSM)* equal the observed marginal means. Our study design is highly unbalanced and thus the *LSM* of the response variable for any effect level will not coincide with the simple within-effect level mean response.

Each least-squares mean is computed as $L'\hat{\beta}$ for a given vector L. For example, in a model with two factors A and B, where A has three levels and B has two levels, the least squares mean response for the first level of factor A is given by:

$$LSM(A_i) = L'\hat{\beta} = \left[1\,1\,0\,0\,\frac{1}{2}\,\frac{1}{2}\right]\hat{\beta},$$

where the first coefficient 1 in L corresponds to the intercept, the next three coefficients correspond to the three levels of factor A and the last two coefficients correspond to the two levels of factor B. If the model also includes an interaction between A and B, then L and $\hat{\beta}$ has an additional 3×2 elements. The corresponding values of the additional six elements in L would be $\frac{1}{2}$ for the two interaction levels involving the first level of factor A (A_1B_1, A_1B_2) and 0 for the four interaction levels that do not involve the first level of factor A $(A_2B_1, A_sB_2, A_3B_1, A_3B_2)$. The coefficient vector L is constructed in a similar way to compute the *LSM* of y (or a transformation of y) for the remaining two levels of A, two levels of B, and even for the six levels of the interaction between A and B if it is present in the model.

When the response variable has been transformed prior to fitting the model, the *LSM* is computed in the transformed scale and must be then transformed back into the original scale. If we have MLEs of the regression coefficients, we can easily compute the *LSMs* in the original scale simply by applying the inverse transformation to $L'\hat{\beta}$. For example, if $g(\mu) = \log(\mu) = x\beta$ and $L'\hat{\beta}$ is the least squares mean in the transformed scale, we can compute the *LSM* in the original scale as

$$LSM_{original} = g^{-1}\left(LSM_{transformed}\right) = g^{-1}\left(L'\hat{B}\right) = \exp\left(L'\hat{B}\right)$$

If the transformation was the logit transformation, the *LSM* in the original scale is computed as

$$LSM_{original} = g^{-1}\left(LSM_{transformed}\right) = g^{-1}\left(L'\hat{B}\right) = \frac{\exp\left(L'\hat{B}\right)}{1 + \exp\left(L'\hat{B}\right)}.$$

VARIANCE OF A NONLINEAR FUNCTION OF PARAMETERS

Suppose that we fit a model to a response variable that has been transformed using some function g as above, and obtain an estimate of a mean $L'\hat{\beta}$. Programs including SAS will also output an estimate of the variance of $L'\hat{\beta}$. We can compute the estimate of the mean in the original scale by applying the inverse transformation g^{-1} to $L'\hat{\beta}$ as described above. In order to obtain an estimate of the variance of $g^{-1}\left(L'\hat{\beta}\right)$, however, we need to make use of, for example, the Delta method, which we now explain.

Given any non-linear function H of some scalar-valued random variable θ, $H(\theta)$ and given σ^2, the variance of θ, we can obtain an expression for the variance of $H(\theta)$ as follows:

$$Var(H(\theta)) = \left[\frac{\partial H(\theta)}{\partial \theta}\right]^2 \sigma^2.$$

For example, suppose that we used a log transformation on a response variable and obtained an *LSM* in the transformed scale that we denote $L'\hat{\beta}$, with estimated variance $\hat{\sigma}_{L'\hat{\beta}}$. The estimate of the mean in the original scale is obtained by applying the inverse transformation to the *LSM*:

$$\hat{m} = LSM_{original} = \exp\left(L'\hat{\beta}\right)$$

The variance of $\hat{m}$ is given by:

$$\hat{\sigma}_{\hat{m}}^2 = \left[\frac{\partial \exp\left(L'\hat{\beta}\right)}{\partial L'\hat{\beta}}\right]^2 \hat{\sigma}_{L'\hat{\beta}} = \left[\exp\left(L'\hat{\beta}\right)\right]^2 \hat{\sigma}_{L'\hat{\beta}}.$$

Suppose now that the response variable was binary and that we used a logit transformation so that

$$g(\mu) = \log\left(\frac{\mu}{1-\mu}\right).$$

Given an MLE $\hat{\beta}$ and an estimate of $L'\hat{\beta}$ the least squares mean in the transformed scale, we compute $\hat{m}$ and $\hat{\sigma}_{\hat{m}}^2$ as follows:

$$\hat{m} = \frac{\exp\left(L'\hat{\beta}\right)}{1+\exp\left(L'\hat{\beta}\right)},$$

$$\hat{\sigma}_{\hat{m}}^2 = \left[\frac{\exp\left(L'\hat{\beta}\right)}{\left[1+\exp\left(L'\hat{\beta}\right)\right]}\right]^2 \hat{\sigma}_{L'\hat{\beta}}.$$

Given a point estimate of the least squares mean in the original scale and an approximation to its variance, we can compute an approximate $100(1-\alpha)\%$ confidence interval for the true mean in the original scale in the usual manner:

$$100(1-\alpha)\% \text{ for } m = \hat{m} \pm t_{df;\alpha/2}\sqrt{\hat{\sigma}_{\hat{m}}^2},$$

where df is the appropriate degrees of freedom. In our case, and due to relatively large sample sizes everywhere, the t critical value can be replaced by the corresponding upper $\alpha/2$ tail of the standard normal distribution.

Appendix 3-8

Main Considerations for Taking a Position by Number of Respondents Saying "Yes"

Consideration	Gender of Respondent	
	Male	Female
Pay	90	88
Benefits	65	62
Promotion opportunities	101	91
Start-up package	131	117
Funding opportunities	96	100
Family-related reasons	120	168
Job location	156	176
Collegiality	170	209
Reputation of department or university	184	224
Quality of research facilities	152	155
Access to research facilities	130	134
Opportunities for research collaboration	179	216
Desire to build or lead a new program or area of research	165	152
This was the only offer I received	52	48

NOTE: There were a total of 612 males and 666 females that responded in each category.

Appendix 4-1
Distribution of Undergraduate Course Load for Faculty by Gender and Discipline

Two statistical tests were carried out. First, a chi-square test of independence of rows was applied to determine whether the *pattern* of the number of undergraduate courses taught[1] by men and women differed. (These tests were either on three or four degrees of freedom.) The tests were not significant at the .05 level except for electrical engineering. It is important to mention that one could have different patterns without having women teach more of fewer courses. For instance, men might teach 1 or 2 courses more often than women do, who in turn might teach 0 or 3 courses more often, but where the mean number of courses remained close.

Therefore, we added a simple two-sample t-test of the average number of courses for men and women. The means are displayed below for each of the disciplinary areas. The t-tests were all not significant at the .05 for the null hypothesis of no difference, again except for electrical engineering. It is clear from the table that men teach more undergraduate courses than do women.

BIOLOGY

Courses Taught	0	1	2	3	4	Total
Men	31	55	12	2	0	100
Women	31	58	11	2	2	104
Total	62	113	23	4	2	204

Chi-squared test of independence: 2.05 (4 degrees of freedom), p-value 0.73.
Means: Men .85 vs. Women .90, t-test is equal to -0.51 p-value 0.61.

CHEMISTRY

Courses Taught	0	1	2	3	Total
Men	43	49	8	1	101
Women	43	48	4	2	97
Total	86	97	12	3	198

Chi-squared test of independence: 1.60 (3 degrees of freedom), p-value 0.66.
Means: Men .67 vs. Women .64, t-test is equal to 0.36 p-value 0.72.

---

[1] Fractional courses were rounded up to the nearest integral number of courses. Missing data was removed from the data prior to analysis. Finally, the data were from the committee's survey of faculty.

MATHEMATICS

Courses Taught	0	1	2	3	Total
Men	21	30	15	2	68
Women	22	38	24	0	84
Total	43	68	39	2	152

Chi-squared test of independence: 3.39 (3 degrees of freedom), p-value 0.33.
Means: Men .97 vs. Women 1.02, t-test is equal to -0.42 p-value 0.68.

ELECTRICAL ENGINEERING

Courses Taught	0	1	2	3	Total
Men	33	46	14	1	94
Women	44	41	4	2	91
Total	77	87	18	3	185

Chi-squared test of independence: 7.70 (3 degrees of freedom), p-value 0.05.
Means: Men .82 vs. Women .60, t-test is equal to 2.09 p-value 0.04.

PHYSICS

Courses Taught	0	1	2	3	Total
Men	33	53	9	0	95
Women	31	66	14	1	112
Total	64	119	23	1	207

Chi-squared test of independence: 2.19 (3 degrees of freedom), p-value 0.53.
Means: Men .75 vs. Women .87, t-test is equal to -1.34 p-value 0.18.

CIVIL ENGINEERING

Courses Taught	0	1	2	3	4	Total
Men	22	44	13	4	0	83
Women	36	67	13	3	1	120
Total	58	111	26	7	1	203

Chi-squared test of independence: 2.63 (4 degrees of freedom), p-value 0.62.
Means: Men .99 vs. Women .88, t-test is equal to 0.94 p-value 0.35.

Appendix 4-2

Percentage of Faculty Members Who Do No Graduate Teaching

Discipline	Men	Women
Chemistry	42.0 (100)	37.5 (96)
Mathematics	63.4 (82)	56.0 (116)
Electrical engineering	55.3 (94)	48.9 (90)
Physics	55.9 (68)	44.6 (83)
Civil engineering	35.1 (97)	22.3 (112)

NOTE: Numbers in parentheses are the total number of respondents in each category.
SOURCE: Survey of faculty conducted by the Committee on Gender Differences in Careers of Science, Engineering, and Mathematics Faculty.

Appendix 4-3

Percentage of Faculty Members Receiving a Reduced Teaching Load When Hired

Discipline	Men	Women
Chemistry	76.9 (52)	80.8 (52)
Mathematics	75.6 (41)	87.0 (69)
Electrical engineering	82.7 (52)	85.5 (55)
Physics	64.5 (31)	71.7 (46)
Civil engineering	70.0 (50)	75.3 (69)

NOTE: Numbers in parentheses are the total number of respondents in each category.
SOURCE: Survey of faculty conducted by the Committee on Gender Differences in Careers of Science, Engineering, and Mathematics Faculty.

Appendix 4-4

Percentage of Faculty Members Who Served on an Undergraduate Thesis or Honors Committee

Discipline	Men	Women
Biology	36.6 (93)	45.3 (96)
Chemistry	26.0 (77)	30.4 (79)
Mathematics	15.4 (65)	13.0 (92)
Electrical engineering	36.6 (93)	45.3 (86)
Physics	26.0 (77)	30.4 (79)
Civil engineering	15.4 (65)	13.0 (92)

NOTE: Numbers in parentheses are the total number of respondents in each category.
SOURCE: Survey of faculty conducted by the Committee on Gender Differences in Careers of Science, Engineering, and Mathematics Faculty.

Appendix 4-5

Percentage of Faculty Members Who Served on and Chaired an Undergraduate Thesis or Honors Committee

Discipline	Men		Women	
	Served	Chair	Served	Chair
Biology	62.30 (38)	27.30 (23)	59.52 (50)	40.78 (34)
Chemistry	57.14 (20)	42.86 (15)	46.15 (24)	53.85 (28)
Mathematics	85.71 (6)	14.29 (1)	30.00 (3)	70.00 (7)
Electrical engineering	43.59 (17)	56.41 (22)	50.00 (17)	50.00 (17)
Physics	62.50 (20)	27.50 (12)	54.54 (18)	45.46 (15)
Civil engineering	62.50 (10)	27.50 (6)	42.86 (12)	57.14 (16)

NOTE: Numbers in parentheses are the total number of respondents in each category.
SOURCE: Survey of faculty conducted by the Committee on Gender Differences in Careers of Science, Engineering, and Mathematics Faculty.

Appendix 4-6

Distribution of Number of Graduate Thesis or Honors Committees for
Research I Tenure and Tenure-Track Faculty: *Men/Women*

Discipline	0	1-3	4-5	6-10	11-30	Total
Biology	9.3	34.3	24.1	22.2	10.2	108
	5.1	41.5	12.7	26.3	14.4	118
Chemistry	6.5	32.7	19.6	23.4	17.8	107
	6.0	39.0	15.0	25.0	15.0	100
Mathematics	43.7	47.9	5.6	2.8	0	71
	35.6	49.4	10.3	3.4	1.2	87
Electrical engineering	11.0	54.0	19.0	12.0	4.0	99
	19.0	37.0	25.0	16.0	3.0	100
Physics	15.4	61.5	16.4	5.8	1.0	104
	29.9	50.4	17.1	2.6	0	117
Civil engineering	4.8	54.8	23.8	10.7	5.9	84
	11.8	18.4	22.2	19.6	14.3	113

NOTE: These are percentages of men and women who fall into each category.
SOURCE: Survey of faculty conducted by the Committee on Gender Differences in Careers of Science, Engineering, and Mathematics Faculty.

Appendix 4-7

Percentage of Time Spent in Administration or Committee Work on Campus and Service to the Profession Outside the University for Tenured and Tenure-Track Faculty at Research I Institutions: *Men/Women*

Discipline	Mean Hours (standard deviation, sample size)
Biology	13.1 (10.7, 110)
	15.6 (11.7, 117)
Chemistry	14.6 (12.5, 108)
	14.8 (10.7, 96)
Mathematics	12.7 (14.3, 81)
	13.6 (11.0, 82)
Electrical engineering	12.9 (11.3, 101)
	17.6 (16.3, 102)
Physics	13.8 (11.5, 108)
	13.9 (12.6, 119)
Civil engineering	19.3 (17.9, 85)
	17.1 (13.5, 116)

SOURCE: Survey of faculty conducted by the Committee on Gender Differences in Careers of Science, Engineering, and Mathematics Faculty.

Appendix 4-8

Distribution of Number of Service Committees for Research I Tenure and
Tenure-Track Faculty: *Men/Women*

Discipline	0	1	2	3	4	5	>5	Total
Biology	34.1	21.4	15.9	13.5	7.9	5.6	1.6	126
	25.6	18.8	16.5	18.8	14.3	1.5	4.5	133
Chemistry	30.3	16.4	23.8	18.0	4.9	4.9	1.6	122
	25.0	24.1	17.0	18.7	8.0	5.4	1.8	112
Mathematics	49.5	20.2	11.1	9.1	7.1	1.0	2.0	99
	41.7	17.5	15.5	13.6	6.8	4.9	0.0	103
Electrical engineering	36.2	28.4	17.2	6.9	6.0	4.3	0.9	116
	34.5	21.6	17.2	15.5	6.0	5.2	0.0	116
Physics	24.0	21.5	22.3	18.2	5.0	3.3	5.8	121
	30.9	30.1	22.8	8.8	2.9	1.5	2.9	136
Civil engineering	28.9	18.6	22.7	9.3	15.5	2.1	3.1	97
	21.9	10.6	25.2	18.7	5.7	9.8	8.1	123

SOURCE: Departmental survey conducted by the Committee on Gender Differences in Careers of Science, Engineering, and Mathematics Faculty.

Appendix 4-9

Mean Salary by Gender and Professorial Rank for Tenure and Tenure-Track
Faculty in Research I Institutions

Discipline	Rank	Mean (1000s) Men	Mean (1000s) Women
Biology	1	101.9 (34)	93.5 (34)
Chemistry	1	112.9 (43)	101.7 (28)
Mathematics	1	106.5 (40)	101.1 (26)
Electrical engineering	1	107.9 (27)	110.2 (33)
Physics	1	110.0 (48)	93.7 (33)
Civil engineering	1	115.0 (24)	102.5 (26)
Biology	2	72.8 (31)	68.2 (48)
Chemistry	2	72.9 (28)	72.7 (36)
Mathematics	2	68.1 (17)	69.0 (29)
Electrical engineering	2	83.8 (25)	93.5 (34)
Physics	2	73.2 (31)	74.8 (34)
Civil engineering	2	81.8 (30)	81.3 (42)
Biology	3	62.2 (35)	59.5 (26)
Chemistry	3	59.6 (33)	62.9 (30)
Mathematics	3	61.1 (22)	58.4 (32)
Electrical engineering	3	76.6 (43)	76.2 (30)
Physics	3	65.1 (26)	65.0 (44)
Civil engineering	3	71.1 (25)	68.9 (42)

NOTES: Rank is denoted as full (1), associate (2), or assistant (3) professor. Salaries are expressed as number of thousands of dollars with number of respondents in parentheses. For example, 34 men at the full professor rank in biology responded that they earn an average of $101.900 per year. Of 1,404 full-time faculty members who responded, only 1,179 included salary data. The salaries expressed are 9-month salaries. Some clearly high outliers were removed. Twenty percent of the respondents replied back with salaries below $100 for 9 months. Since it was likely that these values were actually in the thousands, these numbers were multiplied by 1,000 for the final value rather than omitting the information.

SOURCE: Survey of faculty conducted by the Committee on Gender Differences in Careers of Science, Engineering, and Mathematics Faculty.

Appendix 4-10

Percentage of Tenured and Tenure-Track Faculty in Research I Institutions Receiving Summer Support

Discipline	Men	Women
Biology	66.1 (62)	63.3 (49)
Chemistry	71.2 (59)	81.8 (55)
Mathematics	42.9 (35)	29.1 (55)
Electrical engineering	85.2 (61)	77.4 (62)
Physics	71.4 (49)	63.5 (74)
Civil engineering	80.0 (45)	85.5 (69)

NOTES: Only one-half of those surveyed responded to this question. Numbers in parentheses are the total number of respondents in each category. For example, 66.1 percent of men in biology out of 62 respondents received summer salary support.

SOURCE: Survey of faculty conducted by the Committee on Gender Differences in Careers of Science, Engineering, and Mathematics Faculty.

Appendix 4-11

Percentage of Tenured and Tenure-Track Faculty in Research I Institutions Receiving Travel Funds

Discipline	Men	Women
Biology	45.2 (62)	44.9 (49)
Chemistry	50.8 (59)	32.7 (55)
Mathematics	62.9 (35)	80.0 (55)
Electrical engineering	62.3 (61)	53.2 (62)
Physics	59.2 (49)	64.9 (74)
Civil engineering	64.4 (45)	71.0 (69)

NOTES: Only one-half of those surveyed responded to this question. The numbers are expressed in percentages of the total respondents (parentheses). For example, 45.2 percent of male faculty in biology out of a total of 62 respondents said they had received travel support.

SOURCE: Survey of faculty conducted by the Committee on Gender Differences in Careers of Science, Engineering, and Mathematics Faculty.

Appendix 4-12

Median Square Footage of Lab Space of Faculty Who Report Doing
Experimental Work

Discipline	Men	Women
Biology	1200 (97)	1050 (106)
Chemistry	1500 (94)	1500 (88)
Mathematics	[a]	[a]
Electrical engineering	550 (50)	450 (53)
Physics	1079 (55)	800 (59)
Civil engineering	738 (50)	800 (64)

[a]Mathematics was excluded from this analysis because the small sample size was inadequate for analysis and ran the risk of potentially violating confidentiality.

NOTE: The median square footage of lab space given to faculty members that identify at least some of their research as "experimental." The number of respondents is in parentheses.

SOURCE: Survey of faculty conducted by the Committee on Gender Differences in Careers of Science, Engineering, and Mathematics Faculty.

Appendix 4-13

Faculty Who Have Received More Lab Space Since Hire (Values Are Percentages)

Discipline	Men		Women	
	%	(n/total)	%	(n/total)
Biology	0.25	14/56	0.28	13/47
Chemistry	0.43	23/54	0.49	23/47
Mathematics		[a]		[a]
Electrical engineering	0.24	8/34	0.26	10/38
Physics	0.24	7/29	0.29	12/42
Civil engineering	0.16	5/31	0.14	5/35
Overall	0.28	57/204	0.30	63/209

[a]Mathematics was excluded from this analysis because the small sample size was inadequate for analysis and ran the risk of potentially violating confidentiality.
NOTE: Sample sizes are indicated in parentheses.
SOURCE: Survey of faculty conducted by the Committee on Gender Differences in Careers of Science, Engineering, and Mathematics Faculty.

Appendix 4-14

Percentage of Tenured and Tenure-Track Faculty in Research I Institutions Receiving Sufficient Equipment

Discipline	Men	Women
Biology	99.2 (107)	94.2 (120)
Chemistry	92.5 (107)	88.8 (98)
Mathematics	100.0 (68)	97.0 (66)
Electrical engineering	90.4 (104)	88.8 (98)
Physics	99.0 (103)	97.2 (109)
Civil engineering	87.5 (80)	88.6 (114)

NOTES: The numbers are expressed in percent of total respondents (in parentheses). For example, 99.2 percent of men in biology, out of a total of 107 respondents, stated that they had sufficient equipment.
SOURCE: Survey of faculty conducted by the Committee on Gender Differences in Careers of Science, Engineering, and Mathematics Faculty.

Appendix 4-15

Number of Postdoctorate Students for Tenured and Tenure-Track Faculty in
Research I Institutions (presented by Men and Women)

Discipline	0	1	2	3	4	>4	Weighted Average	Total
Biology	37.7	25.5	11.3	2.8	7.5	9.4	1.33	106
	51.3	25.2	6.7	2.5	5.0	5.9	0.96	119
Chemistry	46.4	15.2	15.2	8.0	4.5	7.1	1.23	112
	39.6	30.7	12.9	6.9	3.0	3.0	1.04	101
Civil engineering	76.6	11.7	11.7	16.9	0.0	0.0	0.86	77
	76.9	15.4	6.6	12.1	1.1	0.0	0.69	91
Electrical engineering	79.8	19.2	9.6	0.0	0.0	0.0	0.38	104
	72.5	16.7	8.8	0.0	0.0	0.0	0.34	102
Mathematics	42.1	30.8	19.6	0.0	1.9	0.9	0.82	107
	37.8	40.3	16.0	1.7	0.8	0.8	0.85	119
Physics	74.4	20.9	3.5	5.8	0.0	0.0	0.45	86
	74.6	18.6	5.1	4.2	0.8	0.0	0.45	118

NOTES: Numbers are expressed as percentage of total and provide the distribution of the number of postdoctorate students and the number of postdoctorate students for each discipline. The final column ">4" deicts when the number of doctoral students was 5 or greater.

SOURCE: Survey of faculty conducted by the Committee on Gender Differences in Careers of Science, Engineering, and Mathematics Faculty.

Appendix 4-16

Percentage of Tenured and Tenure-Track Faculty in Research I Institutions
Receiving Sufficient Clerical Support

Discipline	Men	Women
Biology	50.0 (92)	41.6 (89)
Chemistry	54.0 (100)	30.7 (88)
Civil engineering	72.2 (72)	46.9 (81)
Electrical engineering	56.2 (89)	47.9 (94)
Mathematics	54.9 (102)	44.6 (112)
Physics	35.3 (68)	28.6 (98)

NOTES: Numbers are expressed as percentage of total respondents (in parentheses). For example, 50 percent of men in biology out of 92 respondents believed that they received sufficient clerical support.
SOURCE: Survey of faculty conducted by the Committee on Gender Differences in Careers of Science, Engineering, and Mathematics Faculty.

Appendix 4-17

Percentage of Faculty Members Stating That They Had a Mentor

Discipline	Men	Women
Biology	53.8 (52)	53.7 (67)
Chemistry	54.5 (55)	60.3 (63)
Mathematics	54.0 (50)	50.0 (88)
Electrical engineering	48.4 (64)	72.9 (59)
Physics	28.2 (39)	51.8 (56)
Civil engineering	49.2 (63)	58.5 (82)

NOTES: Faculty in this table includes tenure-track and tenured faculty. Numbers in parentheses are the total number of respondents in each category.
SOURCE: Survey of faculty conducted by the Committee on Gender Differences in Careers of Science, Engineering, and Mathematics Faculty.

Appendix 4-18

Distribution of the Number of Graduate Students for Tenured and Tenure-Track Faculty in Research I Institutions (presented by Men and Women)

Discipline	Number of Students							Sample Size
	0	1	2	3	4	5	>5	
Biology	6.6	9.4	21.7	24.5	13.2	6.6	17.9	106
	6.7	14.3	14.3	16.0	13.4	5.9	29.4	119
Chemistry	4.5	5.4	8.9	18.8	8.9	7.1	46.4	112
	1.0	5.9	8.9	8.9	16.8	11.9	45.5	101
Mathematics	36.4	30.0	11.7	9.1	9.1	0.0	3.9	77
	38.5	23.1	20.9	4.4	6.6	2.2	4.4	91
Electrical engineering	6.7	9.6	11.5	8.7	9.6	7.7	46.2	104
	4.9	5.9	8.8	10.8	15.7	8.8	45.1	102
Physics	6.5	16.8	18.7	24.3	13.1	10.3	10.3	107
	9.2	20.2	17.6	21.8	11.8	7.6	11.8	119
Civil engineering	5.8	4.6	11.6	16.2	17.4	11.6	32.6	86
	2.5	7.6	6.8	12.7	11.0	11.9	47.5	118

NOTE: Final column of ">5" depicts 6 or greater graduate students.
SOURCE: Survey of faculty conducted by the Committee on Gender Differences in Careers of Science, Engineering, and Mathematics Faculty.

Appendix 4-19

Mean Number of Articles Published in Refereed Journals (sole and co-authored) Over the Past 3 Years for Tenured and Tenure-Track Faculty in Research I Institutions

Discipline	Men	Women
Biology	6.7 (81)	6.2 (81)
Chemistry	15.8 (89)	9.4 (79)
Civil engineering	5.3 (69)	4.5 (79)
Electrical engineering	5.8 (102)	7.5 (98)
Mathematics	12.4 (94)	10.4 (106)
Physics	7.6 (85)	6.3 (109)

NOTES: Numbers in parentheses are the total number of respondents in each category.

SOURCE: Survey of faculty conducted by the Committee on Gender Differences in Careers of Science, Engineering, and Mathematics Faculty.

Appendix 4-20a

Estimated Probability of Having Grant Funding by Discipline, Gender, and Whether the Faculty Member Has an Assigned Mentor—Assistant Professors Only

Discipline	Gender	Mentor	Probability of Grant	SD	n
Biology	Male	No	0.91	0.06	12
Biology	Male	Yes	0.87	0.10	15
Biology	Female	No	0.64	0.14	5
Biology	Female	Yes	0.88	0.10	5
Chemistry	Male	No	0.89	0.11	12
Chemistry	Male	Yes	0.96	0.04	9
Chemistry	Female	No	0.77	0.08	6
Chemistry	Female	Yes	0.95	0.04	12
Mathematics	Male	No	0.72	0.19	10
Mathematics	Male	Yes	0.83	0.05	6
Mathematics	Female	No	0.59	0.17	7
Mathematics	Female	Yes	0.91	0.08	12
Electrical engineering	Male	No	0.84	0.16	9
Electrical engineering	Male	Yes	0.87	0.11	21
Electrical engineering	Female	No	0.37	0.09	12
Electrical engineering	Female	Yes	0.86	0.06	12
Physics	Male	No	0.9	0.05	13
Physics	Male	Yes	0.92	0.04	10
Physics	Female	No	0.71	0.18	12
Physics	Female	Yes	0.95	0.03	19
Civil engineering	Male	No	0.87	0.07	10
Civil engineering	Male	Yes	0.53	0.04	12
Civil engineering	Female	No	1	0.00	11
Civil engineering	Female	Yes	1	0.00	12

SOURCE: Survey of faculty conducted by the Committee on Gender Differences in Careers of Science, Engineering, and Mathematics Faculty.

Appendix 4-20b

Estimated Probability of Having Grant Funding by Discipline, Gender, and Whether the Faculty Member Has an Assigned Mentor—Associate Professors Only

Discipline	Gender	Mentor	Probability of Grant	SD	n
Biology	Male	No	0.91	0.09	12
Biology	Male	Yes	0.93	0.12	5
Biology	Female	No	0.83	0.12	23
Biology	Female	Yes	0.97	0.02	10
Chemistry	Male	No	0.94	0.07	13
Chemistry	Male	Yes	0.95	0.07	8
Chemistry	Female	No	0.89	0.09	12
Chemistry	Female	Yes	0.98	0.01	7
Mathematics	Male	No	0.54	0.23	9
Mathematics	Male	Yes	0.74	0.07	2
Mathematics	Female	No	0.85	0.1	12
Mathematics	Female	Yes	0.98	0.03	3
Electrical engineering	Male	No	0.88	0.05	9
Electrical engineering	Male	Yes	0.87	0.09	3
Electrical engineering	Female	No	0.7	0.14	10
Electrical engineering	Female	Yes	0.95	0.02	10
Physics	Male	No	0.93	0.05	11
Physics	Male	Yes	0.96	0.05	10
Physics	Female	No	0.88	0.08	13
Physics	Female	Yes	0.98	0.02	17
Civil engineering	Male	No	0.87	0.09	12
Civil engineering	Male	Yes	0.95	0.02	8
Civil engineering	Female	No	1	0	16
Civil engineering	Female	Yes	1	0	11

SOURCE: Survey of faculty conducted by the Committee on Gender Differences in Careers of Science, Engineering, and Mathematics Faculty.

Appendix 4-21

Percentage of Faculty Missing Salary Data by Gender and Discipline

Discipline	Men	Women
Biology	20.6 (126)	17.3 (133)
Chemistry	13.9 (122)	15.2 (112)
Civil engineering	20.2 (99)	15.5 (103)
Electrical engineering	18.1 (116)	16.4 (116)
Mathematics	12.4 (121)	15.4 (136)
Physics	17.5 (97)	10.6 (123)

NIOTES: Number in parentheses are the total number of respondents in each category. For example, 20.6 percent of men in biology out of 126 total respondents were missing salary data.
SOURCE: Survey of faculty conducted by the Committee on Gender Differences in Careers of Science, Engineering, and Mathematics Faculty.

Appendix 4-22

Percentage of Tenured and Tenure-Track Faculty at Research I Institutions That Were Nominated for at Least One Award

Discipline	Men	Women
Biology	28.4 (109)	15.1 (119)
Chemistry	39.0 (113)	38.6 (101)
Mathematics	31.7 (82)	18.3 (93)
Electrical engineering	19.0 (105)	25.2 (103)
Physics	32.4 (108)	35.2 (122)
Civil engineering	15.3 (85)	23.9 (117)

NOTES: Number in parentheses are the total number of respondents in each category. For example, 28.4 percent of men in biology out of 109 total respondents have been nominated for an award.
SOURCE: Survey of faculty conducted by the Committee on Gender Differences in Careers of Science, Engineering, and Mathematics Faculty.

Appendix 4-23

Percentage of Tenured and Tenure-Track Research I Faculty with Offers to Leave

Discipline	Men	Women
Biology	24.1 (79)	22.3 (94)
Chemistry	38.8 (85)	25.7 (74)
Mathematics	18.1 (72)	33.3 (66)
Electrical engineering	29.3 (58)	46.6 (73)
Physics	36.5 (85)	28.6 (84)
Civil engineering	47.8 (67)	43.8 (73)

NOTES: Number in parentheses are the total number of respondents in each category. For example, 24.1 percent of men in biology out of 79 respondents stated that they had received offers from other universities.

SOURCE: Survey of faculty conducted by the Committee on Gender Differences in Careers of Science, Engineering, and Mathematics Faculty.

Appendix 4-24

Percentage of Tenured and Tenure-Track Faculty at Research I Institutions Planning to Leave or Retire

Discipline	Men	Women
Biology	43.1	45.8
Chemistry	31.3	41.0
Mathematics	45.6	41.2
Electrical engineering	29.0	26.5
Physics	31.5	28.3
Civil engineering	36.5	42.5

SOURCE: Survey of faculty conducted by the Committee on Gender Differences in Careers of Science, Engineering, and Mathematics Faculty.

Appendix 5-1
Knowledge of Tenure Procedures by Gender, Rank, and Presence of a Mentor

Presence of a Mentor by Gender and Rank

Rank	Gender	
	Men	Women
Professor	19 (279)	28 (233)
Associate professor	55 (194)	93 (255)
Assistant professor	108 (208)	142 (235)

NOTES: Sample sizes are in parentheses. For example, of 279 respondents, 19 male full professors stated that they had a mentor at some point in their careers.
SOURCE: Survey of faculty conducted by the Committee on Gender Differences in Careers of Science, Engineering, and Mathematics Faculty.

Knowledge of Institutional Tenure Policies by Gender and Presence of a Mentor

Response	Men		Women	
	Mentor	No Mentor	Mentor	No Mentor
No institutional tenure policy present	3	2	2	4
Tenure policy present but not known	30	39	27	42
Knows institution's tenure policies	136	387	221	357

NOTES: A total of 84 men (13 with mentors) and 70 women (13 with mentors) chose not to respond to this question.
SOURCE: Survey of faculty conducted by the Committee on Gender Differences in Careers of Science, Engineering, and Mathematics Faculty.

Knowledge of Institutional Promotion Policies by Gender and Rank

	Men			Women		
Response	Professor	Assoc. Professor	Asst. Professor	Professor	Assoc. Professor	Asst. Professor
No institutional promotion policy present	1	1	3	3	4	3
Promotion policy present but not known	16	29	71	12	68	90
Knows institution's promotion policies	221	141	115	164	158	130

NOTES: A total of 83 men (41 professors, 23 associate professors, and 19 assistant professors) and 71 women (34 professors, 25 associate professors, and 12 assistant professors) chose not to respond to this.

SOURCE: Survey of faculty conducted by the Committee on Gender Differences in Careers of Science, Engineering, and Mathematics Faculty.

Appendix 5-2

Detailed Tenure Information from Departmental Survey

	Men			Women		
	Tenured	Not tenured	Total	Tenured	Not tenured	Total
Biology	89	16	105	29	5	34
Chemistry	79	22	101	11	0	11
Civil engineering	74	15	89	11	2	13
Electrical engineering	91	10	101	9	0	9
Mathematics	106	16	122	14	1	15
Physics	106	7	113	5	0	5
High-prestige institution	79	22	101	11	1	13
Medium-prestige institution	74	12	86	15	0	15
Low-prestige institution	392	52	444	60	7	67
Total	**545**	**86**	**631**			**95**
Public institution	425	54	479	62	5	67
Private institution	130	32	162	17	3	20
Total	**555**	**86**	**641**			**81**
Stop-the-tenure-clock policy	113	22	135	16	1	17
No stop-the-tenure-clock policy	417	60	477	60	6	66
Total	**530**	**82**	**612**			**83**

NOTES: There were 755 tenure decisions reported by 319 departments that reported having at least 1 tenure case during the 2 years of the study. In 631 of those tenure decisions, the candidate was a man. In 124 decisions, the candidate was a woman. We deleted 37 cases in which the candidate was a woman but the department reported having no female tenure-track faculty at the assistant or associate professor levels. Thus there are only 87 tenure decisions involving women. The column labeled Tenured shows the number of decisions that were positive, while the column labeled Not tenured shows the number of negative decisions. There were five decisions for which information about the stop-the-tenure-clock policy was missing that involved women and 19 decisions that involved men.

SOURCE: Departmental surveys conducted by the Committee on Gender Differences in Careers of Science, Engineering, and Mathematics Faculty.

Appendix 5-3

Time Spent in Both Assistant and Associate Professorships

	Number Tenured	Number of Cases
Percent women among tenure-track faculty		
0 – 10	3	3
10.1 – 25	32	32
25.1 – 50	30	35
50.1 – 75	10	13
75.1 – 100	3	3
Percent women among all faculty		
0 – 10	14	14
10.1 – 25	51	55
25.1 – 50	13	17

NOTES: The percentage of women in the tenure pool was computed as the total number of women on tenure-track (both assistant and associate) divided by the total number of tenure-track faculty (both assistant and associate). The percentage of women among all faculty was computed as the total number of women of all ranks, tenured or tenure-track, divided by the total number of faculty of all ranks, tenured or tenure-track. Again, we did not consider the 37 tenure decisions involving a woman where the number of tenure-track women was reported to be zero.

SOURCE: Departmental survey conducted by the Committee on Gender Differences in Careers of Science, Engineering, and Mathematics Faculty.

Appendix 5-4
Years Between Starting Employment and Achieving
Associate Professor Status, by Gender

Percentage breakdown of the number of years between associate professor rank achieved and first faculty or instructional staff by gender, for full-time faculty at Research I institutions with instructional duties for credit, teaching biology, physical sciences, engineering, mathematics or computer science, fall 2003.

	Years Between Achieved Associate Professor and First Started Employment at Postsecondary Institution				
	0	1-5	6-10	11-15	16 or more
Estimates					
Total	3.9 (1.43)	18.9 (2.72)	60.1 (3.11)	9.8 (2.08)	7.4 (1.60)
Men	4.1 (1.61)	20.0 (2.77)	60.3 (3.45)	8.4 (2.02)	7.1 (1.77)
Women	#	13.1 (9.38)	58.7 (9.8)	16.8 (6.92)	8.7 (4.23)

NOTE: Numbers in parentheses represent standard errors of each mean.
— Too few cases to provide a reliable estimate
SOURCE: National Center for Education Statistics (NCES), National Study of Postsecondary Faculty (NSOPF):2004 National Study of Postsecondary Faculty, March 30, 2006.

Appendix 5-5
Years Between Starting Employment and Achieving Full Professor Status, by Gender

Percentage breakdown of the number of years between full professor rank achieved and first faculty or instructional staff by gender, for full-time faculty at Research I institutions with instructional duties for credit, teaching biology, physical sciences, engineering, mathematics or computer science, fall 2003.

	Years Between Achieved Full Professor and First Started Employment at Postsecondary Institution				
	0 years	1-5	6-10	11-15	16 or more
Estimates					
Total	7.8 (1.13)	6.2 (1.21)	35.4 (2.54)	39.3 (2.33)	11.3 (1.29)
Men	8.4 (1.22)	6.7 (1.31)	36.3 (2.68)	39.4 (2.47)	9.3 (1.16)
Women	#	#	26.4 (7.54)	38.6 (6.26)	31.3 (7.61)

NOTE: Numbers in parentheses represent standard errors of each mean.
— Too few cases to provide a reliable estimate
SOURCE: National Center for Education Statistics (NCES), National Study of Postsecondary Faculty (NSOPF):2004 National Study of Postsecondary Faculty, March 30, 2006

Appendix 5-6

Patterns of Nonresponse for Tenure Decisions

Field	Departments Reporting Tenure Cases	Departments Reporting No Cases	Responding Departments	Departments Surveyed
Biology	59	17	76	87
Chemistry	58	18	76	87
Civil engineering	46	9	55	69
Electrical engineering	44	15	59	77
Mathematics	57	17	74	86
Physics	60	17	77	86
Total	324	93	417	492

SOURCE: Departmental survey conducted by the Committee on Gender Differences in Careers of Science, Engineering, and Mathematics Faculty.

Appendix 5-7

Patterns of Nonresponse for Promotion Decisions

Field	Departments Reporting Promotion Cases	Departments Reporting No Cases	Responding Departments	Departments Surveyed
Biology	42	31	73	87
Chemistry	68	6	74	87
Civil engineering	41	14	55	69
Electrical engineering	43	16	59	77
Mathematics	46	27	73	86
Physics	49	28	77	86
Total	289	122	411	492

SOURCE: Departmental survey conducted by the Committee on Gender Differences in Careers of Science, Engineering, and Mathematics Faculty.

Bibliography

Ahern, N., and E. Scott. 1981. *Career Outcomes in a Matched Sample of Men and Women Ph.D.s.* Washington, DC: National Research Council.

Allen, T. D., L. T. Eby, S. S. Douthitt, and C. L. Noble. 2002. Applicant gender and family structure: Effects on perceived relocation commitment and spouse resistance. *Sex Roles* 47(11/12):543-552.

Allison, P. 1984. *Event History Analysis: Regression for Longitudinal Event Data.* Newbury Park, CA: Sage Publications.

Alvarez, R. M., and Brehm, J. 1995. American ambivalence towards abortion policy: Development of a heteroskedatic probit model of competing values. *American Journal of Political Science* (39):1055-1089.

Amato, I. 1992. Profile of a field: Chemistry. *Science* 255(5050):1372-1373.

American Association of University Professors (AAUP). 2003. Contingent appointments and the academic profession. *Academe* 89(5):59-69.

American College of Physicians. 1991. Promotion and tenure of women and minorities on medical school faculties. *Annals of Internal Medicine* 114(1):63-68.

Amey, M. J. 1996. The institutional marketplace and faculty attrition. *Thought & Action: The NEA Higher Education Journal* 12:23-35.

Andersen, K., and E. D. Miller. 1997. Gender and student evaluations of teaching. *PS: Political Science and Politics* 30(2):216-219.

Anderson, E. 2002. *The New Professoriate: Characteristics, Contributions, and Compensation.* Washington, DC: American Council on Education.

Antonio, A. 2003. Diverse student bodies, diverse faculties. *Academe* 89(6):14-17.

Aper, J., and J. Fry. 2003. Post-tenure review at graduate institutions in the United States. *The Journal of Higher Education* 74(3):241-260.

Ash, A. S., P. L. Carr, R. Goldstein, and R. H. Friedman. 2004. Compensation and advancement of women in academic medicine: Is there equity? *Annals of Internal Medicine* 141(3):205-212.

Ashenfelter, O., and D. Card. 2002. Did the elimination of mandatory retirement affect faculty retirement flows? *American Economic Review* 92(4):957-980.

Ashraf, J. 1996. The influence of gender on faculty salaries in the United States, 1969-89. *Applied Economics* 28:857-864.

Alvarez, R. M., and Brehm, J. 1995. American ambivalence towards abortion policy: Development of a heteroskedatic probit model of competing values. *American Journal of Political Science* (39):1055-1089.

Association of American Universities (AAU). 1998. *Committee on Postdoctoral Education Report and Recommendations*, March 31. Washington, DC: AAU.

Astin, H., and C. Cress. 2003. A national profile of academic women in research universities. Pp. 53-90 in *Equal Rites, Unequal Outcomes: Women in American Research Universities*, L. Hornig, ed., New York: Kluwer Academic Publishers.

August, L. 2006. Attrition among female tenure-track faculty. Paper presented at the meeting of the Association for Institutional Research, Chicago, May 18.

August, L., and J. Waltman. 2004. Culture, climate, and contribution: Career satisfaction among female faculty. *Research in Higher Education* 45(2):177-192.

Bagihole, B. 1993. How to keep a good woman down: An investigation of the role of institutional factors in the process of discrimination against women academics. *British Journal of Sociology of Education* 14(3):261-274.

Bahrami, B. 2001. Factors affecting faculty retirement decisions. *The Social Science Journal* 38(2):297-305.

Bailyn, L. 2003. Academic careers and gender equity: Lessons learned from MIT. *Gender, Work and Organization* 10(2):137-153.

Bain, O., and W. Cummings. 2000. Academe's glass ceiling: Societal, professional-organizational, and institutional barriers to the career advancement of academic women. *Comparative Education Review* 44(4):493-514.

Barbezat, D. A. 2004. A loyalty tax? National measures of academic salary compression. *Research in Higher Education* 45(7):761-776.

Barbezat, D. A. 2003. Gender Differences in Career Development: A Cohort Study of Economists (January). Available at SSRN: http://ssrn.com/abstract=383822.

Barbezat, D. A. 2003. From here to seniority: The effect of experience and job tenure on faculty salaries. *New Directions for Institutional Research* 117:21-47.

Barbezat, D. A. 2002. History of pay equity studies. *New Directions for Institutional Research* 115:9-39.

Barbezat, D. A. 1992. The market for new Ph.D. economists. *Journal of Economic Education* 23(3):262-276.

Barbezat, D. A. 1991. Updating estimates of male-female salary differentials in the academic labor market. *Economics Letters* 36:191-195.

Barbezat, D. A. 1989. Affirmative action in higher education: Have two decades altered salary differentials by sex and race? *Research in Labor Economics* 10:107-156.

Barbezat, D. A., and J. Hughes. 2005. Salary structure effects and the gender pay gap in academia. *Research in Higher Education* 46(6):621-640.

Barbezat, D. A., and J. Hughes. 2001. The effect of job mobility on academic salaries. *Contemporary Economic Policy* 19(4):409-423.

Barinaga, M. 1992. Profile of a field: Neuroscience. *Science* 255(5050):1366-1367.

Barnes, L. L. B., M. O. Agago, and W. T. Coombs. 1998. Effects of job-related stress on faculty intention to leave academia. *Research in Higher Education* 39(4):457-469.

Basow, S. A., and N.T. Silberg. 1987. Student evaluations of college professors: Are female and male professors rated differently? *Journal of Educational Psychology* 79:308-314.

Baum, R. 2003. Achieving gender equity in chemistry. *Chemical & Engineering News* 81(15):46-47.

Bayer, A., and J. Dutton. 1977. Career age and research-professional activities of academic scientists: Tests of alternative nonlinear models and some implications for higher education faculty policies. *The Journal of Higher Education* 48(3):259-282.

Becker, W., and Toutkoushian R. 2003. Measuring gender bias in the salaries of tenured faculty members. *New Directions for Institutional Research* 117:5-20.

Becker, W., and R. Toutkoushian. 1995. The measurement and cost of removing unexplained gender differences in faculty salaries. *Economics of Education Review* 14(3):209-220.

Bellas, M. L. 1997. Disciplinary differences in faculty salaries: Does gender bias play a role? *Journal of Higher Education* 68(3):299-321.

Bellas, M. L. 1994. Comparable worth in academia: The effects on faculty salaries of the sex composition and labor-market conditions of academic disciplines. *American Sociological Review* 59(6):807-821.

Bellas, M. L. 1993. Faculty salaries: Still a cost of being female? *Social Science Quarterly* 74(1): 62-75.

Bellas, M. L., and Toutkoushian, R.K. 1999. Faculty time allocations and research productivity: Gender, race, and family effects. *Review of Higher Education* 22(4):367-390.

Bellas, M. L., P. N. Ritchey, and P. Parmer. 2001. Gender differences in the salaries and salary growth rates of university faculty: An exploratory study. *Sociological Perspectives* 44(2):163-187.

Bement, Jr., A. L. 2005. Remarks, setting the agenda for 21st century science. Presented at the meeting of the Council of Scientific Society Presidents, Washington, DC, December 5. Available at http://www.nsf.gov/news/speeches/bement/05/alb051205_societypres.jsp.

Benedict, M. E., and L. Wilder. 1999. Unionization and tenure and rank outcomes in Ohio universities. *Journal of Labor Research* 20(2):185-201.

Benjamin, E. 1999. Disparities in the salaries and appointments of academic women and men. *Academe* 85(1):60-62.

Bennof, R. J. 2004. Federal science and engineering obligations to academic and nonprofit institutions reached record highs in FY 2002. *NSF InfoBrief*, June, (NSF 04-324).

Bentley, R. J., and R. T. Blackburn. 1992. Two decades of gains for female faculty? *Teachers College Record* 93(4):697-709.

Beutel, A. M., and D. J. Nelson. 2005. The gender and race-ethnicity of faculty in top science and engineering research departments. *Journal of Women and Minorities in Science and Engineering* 11(4):389-402.

Bhattacharjee, Y. 2005. Princeton resets family-friendly tenure clock. *Science* 309(5739):1308.

Bhattacharjee, Y. 2004a. Children vs. tenure, *Science Now* (June 21):2.

Bhattacharjee, Y. 2004b. Family matters: Stopping tenure clock may not be enough. *Science* 306(5704): 2031, 2033.

Billard, L. 1992. The influence of gender on advancement in astronomy, physics and mathematics. Pp. 23-42 in *Proceedings of Women at Work: A Meeting on the Status of Women in Astronomy*, C. M. Urry, L. Danly, L. E. Sherbert, and S. Gonzaga, eds. Baltimore, MD: Space Telescope Science Institute, September 8-9.

Billard, L. 1991. The past, present, and future of academic women in the mathematical sciences. *Notices* 38(7):707-714.

Bird, S., J. Litt, and Y. Wang. 2004. Creating status of women reports: Institutional housekeeping as "women's work." *NWSA Journal* 16(1):194-207.

Bleak, J., H. Neiman, C. S. Rule, and C. Trower. 2000. *Faculty Recruitment Study: Statistical Analysis Report*. Cambridge, MA: Harvard Graduate School of Education.

Blum, D. 1991. Environment still hostile to women in academe, new evidence indicates. *Chronicle of Higher Education* 38(7):A2.

Boedeker, A. 2003. Science funds low for startup packages. *The Daily Toreador* (October 14).

Bradburn, E. M., and Sikora, A. C. 2002. Gender and racial/ethnic differences in salary and other characteristics of postsecondary faculty: Fall 1998 (NCES 2002–170). U.S. Department of Education. Washington, DC: National Center for Education Statistics.

Bradley, G. 2004. Contingent faculty and the new academic labor system. *Academe* 90(1):28-31.

Brennan, M. B. 1996. Women chemists considering careers at research activities. *Chemical and Engineering News* 74(June 10): 8-15.

Bronstein, P., and L. Farnsworth. 1998. Gender differences in faculty experiences of interpersonal climate and processes for advancement. *Research in Higher Education* 39(5):557-585.

Brown, B. W., and S. A. Woodbury. 1995. *Gender Differences in Faculty Turnover*. Kalamazoo, MI: W.E. Upjohn Institute for Employment Research.

Bugeja, M. 2004. The drama behind the job ad. *Chronicle of Higher Education* 51(13):C1-C2.

Burton, L., and L. Parker. 1998. Degrees and occupations in engineering: How much do they diverge? *NSF Issue Brief*, December, (NSF 99-318).

Byrum, A. 2001. Women's place in ranks of academia. *Chemical and Engineering News* 79(40):98-99.

Callister, R. R. 2006. The impact of gender and department climate on job satisfaction and intentions to quit for faculty in science and engineering fields. *Journal of Technology Transfer* 31(3):367-375.

Campbell, K. 2001b. Leaders of 9 universities and 25 women faculty meet at MIT, agree to equity reviews. *MIT News* (January 30).

Campbell, K. 2001. Statement on gender equity in academic science and engineering. *MIT News* (January 30).

Carr, P. L., A. S. Ash, R. H. Friedman, L. Szalacha, R. C. Barnett, A. Palepu, and M. M. Moskowitz. 2000. Faculty perceptions of gender discrimination and sexual harassment in academic medicine. *Annals of Internal Medicine* 132(11):889-896.

Carr, P. L., R. H. Friedman, M. M. Moskowitz, and L. E. Kazis. 1993. Comparing the status of women and men in academic medicine. *Annals of Internal Medicine* 119(9):908-913.

Carter, O., S. Nathisuwan, G. J. Stoddard, and M. A Munger. 2003. Faculty turnover within academic pharmacy departments. *The Annals of Pharmacotherapy* 37:197-201.

Case Western Reserve University. 2003. Resource Equity at Case Western Reserve: Results of Faculty Focus Groups. Available at http://www.case.edu/president/aaction/resourcequity2003.doc. Accessed April 22, 2009.

Cashin, W. E. 1995. Student ratings of teaching: The research revisited. Idea Paper No. 32, Manhatten, KS. Kansas State University, Center for Faculty Evaluation and Development.

Cataldi, E. F., E. M. Bradburn, and M. Fahimi, 2005. 2004 National Study of Postsecondary Faculty (NSOPF:04): Background Characteristics, Work Activities, and Compensation of Instructional Faculty and Staff: Fall 2003 (NCES 2006-176). U.S. Department of Education. Washington, DC: National Center for Education Statistics.

Cataldi, E. F., M. Fahimi, and E. M. Bradburn. 2005. 2004 National Study of Postsecondary Faculty (NSOPF:04) Report on Faculty and Instructional Staff in Fall 2003 (NCES 2005–172). U.S. Department of Education. Washington, DC: National Center for Education Statistics.

Cauble, S., A. Christy, and M. Lima. 2000. Toward plugging the leaky pipeline: Biological and agricultural engineering female faculty in the United States and Canada. *Journal of Women and Minorities in Science and Engineering* 6:229-249.

Center for Research on Learning and Teaching (CRLT), The University of Michigan. Resources on Faculty Mentoring. Available at http://www.crlt.umich.edu/publinks/facement.html. Accessed on April 22, 2009

Centra, J., and N. Gaubatz. 2000. Is there gender bias in student evaluations of teaching? *The Journal of Higher Education* 70(1):17-33.

Chander, R., and Mervis, J. 2001. The bottom line for U.S. life scientists. *Science Magazine*. 294(554): 395.

COACHE (Collaborative on Academic Careers in Higher Education). 2006. 2006 COACHE Survey. Cambridge MA: Harvard University Graduate School of Education. Available at http://www.gse.harvard.edu/~newscholars/downloads/COACHE_Report_20060925.pdf.

Cohoon, J., McGrath, R. Shwalb, and L.-Y. Chen, 2003. Faculty turnover in CS departments. *ACM SIGCSE Bulletin* 35(1):108-112.

Cole, J. R., and H. Zuckerman. 1984. The productivity puzzle: Persistence and change in patterns of publication among men and women scientists. Pp. 217-258 in *Advances in Motivation and Achievement*, vol. 2, P. Maehr and M. W. Steinkamp, eds. Greenwich, CT: JAI Press.

Cole, M. S., H. S. Feild, and W. F. Giles. 2004. Interaction of recruiter and applicant gender in resume evaluation: A field study. *Sex Roles* 51(9/10):597-608.

Collins, J. 2002. May you live in interesting times: Using multidisciplinary and interdisciplinary programs to cope with change in the life sciences. *BioScience* 52(1):75-83.

Commission on Professionals in Science and Technology (CPST). 2006. *Four Decades of STEM Degrees, 1966-2004: "The Devil is in the Details."* STEM Workforce Data Project: Report No. 6. Washington, DC: CPST.

Conley, V. M. 2005. Career paths for women faculty: Evidence from NSOPF:99. *New Directions for Higher Education* 130:25-39.

Connell, M. A., and F. Savage. 2001. Does collegiality count? *Academe* 87(6):37-41.

Conway, D. 1993. Can statistics tell us what we do not want to hear? The case of complex salary structures: Comment. *Statistical Science* 8(2):158-165.

Corley, E. A. 2005. How do career strategies, gender, and work environment affect faculty productivity levels in university-based science centers? *Review of Policy Research* 22(5):637-655.

Cornell University. 2002. Cornell Higher Education Research Institute (CHERI) Survey on Start-up Costs and Laboratory Allocation Rules: Summary of the Findings. Available at http://www.ilr. cornell.edu/cheri/surveys/2002surveyResults.html. Accessed October 7, 2008.

Cox, A. M. 2000. Study shows colleges' dependence on their part-time instructors. *The Chronicle of Higher Education* 47(14):A12. Rochester, NY.

Cox, D.R., and Oaks, D. 1984 *Analysis of Survival Data.* London: Chapman and Hall.

Cullen, D., and G. Luna. 1993. Women mentoring in academe: Addressing the gender gap in higher education. *Gender & Education* 5(2):125-138.

Curtis, J. 2005. Inequities persist for women and non-tenure-track faculty: The annual report on the economic status of the profession 2004-05. *Academe* 91(2):20-30.

Curtis, J. 2004b. Balancing work and family for faculty. Why it's important. *Academe* 90(6):21-23.

Curtis, J. 2004. Introductory remarks to the Committee on Gender Differences in Careers of Science, Engineering, and Mathematics Faculty. The National Academies. Washington, DC. January 30.

Daly, C. J., and J. R. Dee. 2006. Greener pastures: Faculty turnover intent in urban public universities. *The Journal of Higher Education* 77(5):776-803.

Dean, C. 2006. Women in science: The battle moves to the trenches. *New York Times* 156 (December 19):F1-F7.

Devlin, A. S. 1997. Architects: Gender-role and hiring decisions. *Psychological Reports* 81(2): 667-676.

Ding, W. W., F. Murray, and T. E. Stuart. 2006. Gender differences in patenting in the academic life sciences. *Science* 313(5787):665-667.

Dooris, M. J., M. Guidos, and W. L. Miley. 2006. Tenure achievement rates at research universities. Paper presented at the Annual Forum of the Association for Institutional Research. Chicago, May 17.

Dressel, P., and J. Dietrich. 1967. Departmental review and self-study. *The Journal of Higher Education* 38(1):25-37.

Dresselhaus, M.S. 2007. What we learn from the statistics: Challenges to institutions: recruitment, hiring, retention, promotions. Paper presented at the American Physical Society's conference on Gender Equity: Strengthening the Physics Enterprise in Universities and National Laboratories, College Park, MD, May 6-8.

Dundar, H., and D. R. Lewis. 1998. Determinants of research productivity in higher education. *Research in Higher Education* 39(6):607-631.

Dupree, A. 2001. Evaluation of the status of women in astronomy. *STATUS: A Report on Women in Astronomy* (June):1-4.

Ehrenberg, R. G. 2003. Unequal progress: The annual report on the economic status of the profession, 2002-03. *Academe* 89(2):21-33.

Ehrenberg, R. G. 2001. Career's end. *Academe* 87(4):24-30.

Ehrenberg, R. G. 1992. The flow of new doctorates. *Journal of Economic Literature* 30(2):830-875.

Ehrenberg, R. G., and L. Zhang. 2005. The changing nature of faculty employment. In *Recruitment, Retention and Retirement in Higher Education: Building and Managing the Faculty of the Future*, R. L. Clark and J. Ma, eds., Northampton MA: Edwin Elar Publishing.

Ehrenberg, R. G., and M. J. Rizzo. 2004. Financial forces and the future of American higher education. *Academe* 90(4):28-31.

Ellis, R. 2001. *Academic Chemists 2000: A Decade of Change: 1990-2000*. Washington, DC: American Chemical Society.

Etzkowitz, H., C. Kemelgor, and B. Uzzi 2000. *Athena Unbound: The Advancement of Women in Science and Technology*. Cambridge: Cambridge University Press.

Etzkowitz, H., C. Kemelgor, M. Neuschatz, and B. Uzzi. 1994. Barriers to women in academic science and engineering. Pp. 43-67 in *Who Will Do Science? Educating the Next Generation*, W. Pearson Jr. and I. Fechter, eds., Baltimore, MD: Johns Hopkins University Press.

Etzkowitz, H., C. Kemelgor, M. Neuschatz, B. Uzzi, and J. Alonzo. 1994. The paradox of critical mass for women in science. *Science* 266(5182):51-54.

Euben, D. 2001. Pay equity in the academy. *Academe* 87(4):30-36.

Everett, K. G., W. S. Deloach, and S. E. Bressan. 1996. Women in the ranks: Faculty trends in the ACS-approved departments. *Journal of Chemical Education* 73(2):139-141.

Ezorsky, G. 1977. Hiring women faculty. *Philosophy and Public Affairs* 7(1):82-91.

Fairweather, J. S. 2002. The mythologies of faculty productivity: Implications for institutional policy and decision making. *The Journal of Higher Education* 73(1):26-48.

Fairweather, J. S. 1996. *Faculty Work and Public Trust: Restoring the Value of Teaching and Public Service in American Academic Life*. Boston: Allyn and Bacon.

Fairweather, J. S. 1995. Myths and realities of academic labor markets. *Economics of Education Review* 14(2):179-192.

Fairweather, J. S. 1993. Academic values and faculty rewards. *The Review of Higher Education* 17(1): 43-68.

Farber, S. 1977. The earnings and promotion of women faculty: Comment. *American Economic Review* 67(2):199-206.

Feldman, K. A. 1992. College students' views of male and female college teachers: Part I-Evidence from the social laboratory and experiments. *Research in Higher Education* 33:317-375.

Feldman, K. A. 1993. College students, views of male and female college teachers: Part II-Evidence from students' evaluations of their classroom teachers. *Research in Higher Education* 34:151-211.

Ferber, M. A., and J. Loeb. 2002. Issues in conducting an institutional salary-equity study. *New Directions for Institutional Research* 115:41-70.

Ferber, M. A., and C. A. Green. 1982. Traditional or reverse sex discrimination? A case study of a large public university. *Industrial and Labor Relations Review* 35(4):550-564.

Ferreira, M. 2003. Gender issues related to graduate student attrition in two science departments. *International Journal of Science Education* 25(8):969-989.

Finkel, S. K., S. Olswang, and N. She. 1994. Childbirth, tenure, and promotion for women faculty. *Review of Higher Education* 17:259-270.

Finkelstein, M. 2003. The morphing of the American academic profession. *Liberal Education* 89(4):6-15.

Finkelstein, M., and J. Schuster. 2001. Assessing the silent revolution. *AAHE Bulletin* 54(2):3-7.

Fogg, P. 2003a. Family time. *Chronicle of Higher Education* 49(40):A10.

Fogg, P. 2003b. So many committees, so little time. *Chronicle of Higher Education* 50(17):A14.

Fogg, P. 2003c. The gap that won't go away: Women continue to lag behind men in pay; the reasons may have little to do with gender bias. *Chronicle of Higher Education* 49(32):A12.

Foschi, M., L. Lai, and K. Sigerson. 1994. Gender and double standards in the assessment of job applicants. *Social Psychology Quarterly* 57(4):326-339.

Fowler, Jr., F. 1993. *Survey Research Methods*, 2nd ed, Newbury Park, CA: Sage Publications.

Fox, G., A. Schwartz, and K. M. Hart. 2006. Work-family balance and academic advancement in medical schools. *Academic Psychiatry* 30(3):227-234.

Fox, M. F. 2005. Gender, family characteristics, and publication productivity among scientists. *Social Studies of Science* 35(February):131-150.

Fox, M. F. 2003. Gender, faculty, and doctoral education in science and engineering. Pp. 91-110 in *Equal Rites, Unequal Outcomes: Women in American Research Universities*, L. Hornig, ed. New York: Kluwer Academic Publishers.

Fox, M. F. 2001a. Women, science, and academia: Graduate education and careers. *Gender and Society* 15(5):654-666.

Fox, M. F. 2001b. Women in science and engineering: What we know about education and employment. Pp. 25-28 in *Scientists and Engineers for the New Millennium: Renewing the Human Resource*, D. Chubin and W. Pearson, eds., Washington, DC: Commission on Professionals in Science and Technology.

Fox, M. F. 2000. Organizational environments and doctoral degrees awarded to women in science and engineering departments. *Women's Studies Quarterly* 28(1/2):47-61.

Fox, M. F. 1998. Women in science and engineering: Theory, practice and policy programs. *Signs* 42(1):201-223.

Fox, M. F. 1992. Research, teaching, and publication productivity: Mutuality versus competition in academia. *Sociology of Education* 65(4):293-305.

Fox, M. F. 1991. Gender, environmental milieu, and productivity in science. In *The Outer Circle: Women in the Scientific Community*, H. Zuckerman, J. Cole, and J. Bruer, eds., New York: W. W. Norton.

Fox, M. F. 1985. Publication, performance, and reward in science and scholarship. Pp. 255-282 in *Higher Education: Handbook of Theory and Research*, vol. 1. J. C. Smart, ed. New York: Agathon Press.

Fox, M. F. 1983. Publication productivity among scientists: A critical review. *Social Studies of Science* 13:285-305.

Fox, M. F., and C. Colatrella. 2006. Participation, performance, and advancement of women in academic science and engineering: What is at issue and why. *Journal of Technology Transfer* 31(3):377-386.

Fox, M. F., and P. Stephan. 2001. Careers of young scientists: Preferences, prospects and realities by gender and field. *Social Studies of Science* 31(1):109-122.

Frehill, L. M. 2006. Measuring occupational sex segregation of academic science and engineering. *Journal of Technology Transfer* 31(3):345-354.

Fried, L. P., C. A. Francomano, S. M. MacDonald, E. M. Wagner, E. J. Stokes, K. M. Carbone, W. B. Bias, M. M. Newman, and J. D. Stobo. 1996. Career development for women in academic medicine: Multiple interventions in a department of medicine. *JAMA* 276(11):898-905.

Fry, C., and S. Allgood. 2002. The effect of female student participation in the society of women engineers on retention—A study at Baylor University. Paper presented at the 32nd ASEE/IEEE Frontiers in Education Conference, Session F4C, Boston, November 6-9.

Galemore, G. 2003. Title IX and sex discrimination in education: An overview. *CRS Report for Congress*. Updated March 4.

GAO (U.S. Government Accountability Office). 2004. *Women's Participation in the Sciences Has Increased, but Agencies Need to Do More to Ensure Compliance with Title IX.* GAO-04-639. July. Washington, DC: GAO.

Gander, J. P. 1999. Faculty gender effects on academic research and teaching. *Research in Higher Education* 40(2): 171-184.

Gander, J. P. 1997. Gender-based faculty-pay differences in academe: A reduced-form approach. *Journal of Labor Research* 18(3):451-461.

Gandy, K. 2004. Universities must change faculty atmosphere for women and people of color. Remarks at the National Press Club, Washington, DC (January 15).

Gastwirth, J. 1993. Can statistics tell us what we do not want to hear? The case of complex salary structures: Comment. *Statistical Science* 8(2):165-171.

Gender Equity Committee on Academic Climate 2003. *An Assessment of the Academic Climate for Faculty at UCLA.* Los Angeles, CA: University of California at Los Angeles.

Gibbons, A. 1992a. Key issue: Mentoring. *Science* 255(5050):1368-1369.

Gibbons, A. 1992b. Key issue: Tenure. *Science* 255(5050):1386-1388.

Gibbons, A. 1992c. Key issue: Two-career science marriage. *Science* 255(5050):1380-1381.

Gibbons, M. 2007. Engineering by the Numbers. Statistics from 2006 edition of the Profiles of Engineering and Engineering Technology Colleges, available at http://www.asee.org/publications/profiles/upload/2006ProfileEng.pdf.

Gibbons, M. 2006. The Year in Numbers. Statistics from 2005 edition of the Profiles of Engineering and Engineering Technology Colleges, available at http://www.asee.org/publications/profiles/upload/2005ProfileEng.pdf.

Gibbons, M. 2005. The Year in Numbers. Statistics from 2004 edition of the Profiles of Engineering and Engineering Technology Colleges, available at http://www.asee.org/publications/profiles/upload/2004ProfileIntro2.pdf.

Gibbons, M. 2004. A New Look at Engineering. Statistics from 2003 edition of the Profiles of Engineering and Engineering Technology Colleges, available at http://www.asee.org/about/publications/profiles/upload/2003engprofile.pdf.

Gibbons, M. 2003. *2002 Profiles of Engineering and Engineering Technology Colleges.* Washington, DC: ASEE.

Gilbert, J. R., E. S. Williams, and G. D. Lundberg. 1992. Is there gender bias in JAMA's peer review process? *JAMA* 272:139-142.

Ginther, D. 2004. Why women earn less: Economic explanations for the gender salary gap in science. *AWIS Magazine* 33(1):6-10.

Ginther, D. 2003. Is MIT an exception? Gender pay differences in academic science. *Bulletin of Science, Technology & Society* 23(1):21-26.

Ginther, D. 2001. Does Science Discriminate against Women? Evidence from Academia, 1973-97. Federal Reserve Bank of Atlanta Working Paper 2001-02.

Ginther, D., and K. Hayes. 2003. Gender differences in salary and promotion for faculty in the humanities, 1977-95. *Journal of Human Resources* 38:34-73.

Ginther, D., and K. Hayes. 1999. Gender differences in salary and promotion in the humanities. *American Economic Review* 89(2):397-402.

Ginther, D., and S. Kahn. 2004. Women in economics: Moving up or falling off the academic career ladder? *Journal of Economic Perspectives* 18(3):193-214.

Ginther, D., and S. Kahn. 2006. Does Science Promote Women? Evidence from Academia 1973-2001. NBER Working Paper No. W12691 Workshop presentation.

Glover, D., and B. Parsad. 2002. The Gender and Racial/Ethnic Composition of Postsecondary Instructional Faculty and Staff: 1992–98 (NCES 2002–160). U.S. Department of Education. Washington, DC: National Center for Education Statistics.

Golde, C. M. 1998. Beginning graduate school: Explaining first-year doctoral attrition. *New Directions in Higher Education* 101:55-64.

Goldin, C., and C. Rouse. 2000. Orchestrating impartiality: The impact of 'blind' auditions on female musicians. *American Economic Review* 90(4):715-741.

Goldin, D. Q. 2006. Geographic shackles and the academic careers of women. *Computing Research News* 18(2):2, 6.

Gomez-Mejia, L. R., and D. B. Balkin. 1992. Determinants of faculty pay: An agency theory perspective. *Academy of Management Journal* 35(5):921-955.

Gottselig, M., and L. Oeltjen. 2003. Choosing a grad school. *Chemical and Engineering News* 81(15):4.

Grant, L., and K. B. Ward. 1991. Gender and publishing in sociology. *Gender and Society* 5(2):207-223.

Gray, M. 1993. Can statistics tell us what we do not want to hear? The case of complex salary structures. *Statistical Science* 8(2):144-158.

Guillory, E. 2001. The black professoriate: Explaining the salary gap for African-American female professors. *Race Ethnicity and Education* 4(3):129-148.

Gunter, R., and A. Stambach. 2005. Differences in men and women scientists' perceptions of workplace climate. *Journal of Women and Minorities in Science and Engineering* 11(1):97-116.

Gunter, R., and A. Stambach. 2003. As balancing act and as game: How women and men science faculty experience the promotion process. *Gender Issues* 21(1):24-42.

Hagedorn, L. S. 2000. Conceptualizing faculty job satisfaction: Components, theories, and outcomes. *New Directions for Institutional Research* 105:5-20.

Hagedorn, L. S., and L. Sax. 1999. Marriage, children, and aging parents: The role of family-related factors in faculty job satisfaction. Paper presented at the annual conference of the American Educational Research Association. Montreal, Canada, April 19-23.

Hamel M. B., J. R. Ingelfinger, E. Phimister, and C. G. Solomon. 2006. Women in academic medicine—progress and challenges. *New England Journal of Medicine* 355(3):310-312.

Hansen, W. L. 1988. Merit pay in structured and unstructured salary systems. *Academe* 74(6): 10-13.

Hargens, L., and J. S. Long. 2002. Demographic inertia and women's representation among faculty in higher education. *The Journal of Higher Education* 73(4):494-517.

Harms, W. 2001. NSF announces institutional transformation awards under "advance." NSF PR 01-79. NSF Office of Legislative and Public Affairs. Available at http://www.nsf.gov/od/lpa/news/press/01/pr0179.htm.

Harrigan, M. N. 1999. An analysis of faculty turnover at the University of Wisconsin-Madison. Paper presented at the 39th Annual AIR Forum, Seattle, May 30-June 2.

Heilman, M. E. 2001. Description and prescription: How gender stereotypes prevent women's ascent up the organizational ladder. *Journal of Social Issues* 57(4):657-674.

Henderson, A. 2007. Dual-career academic couples. *ASCB Newsletter* 30(2):46-47.

Henry, C. M. 2002. Women welcome: Symposium spotlights departments with positive environments for women. *Chemical & Engineering News* 80(38):106-108.

Hensel, N. 1991. Realizing gender equality in higher education: The need to integrate work/family issues. ASHE-ERIC Higher Education Report No. 2. Washington, DC: The George Washington University, School of Education and Human Development.

Heylin, M. 2007. A long-term revolution. *Chemical & Engineering News* 85(7):67.

Hill, S. 2006. S&E doctorates hit all-time high in 2005. *NSF InfoBrief*, November, (NSF 07-301).

Hill, S., T. Hoffer, and M. Golladay. 2004. Plans for postdoctoral research appointments among recent U.S. doctorate recipients. *NSF InfoBrief*, March, (NSF 04-308).

Hoffer, T. B., L. Selfa, V. Welch, Jr., K. Williams, M. Hess, J. Friedman, S. C. Reyes, K. Webber, and I. Guzman-Barron. 2004. Doctorate Recipients from United States Universities: Summary Report 2003. Chicago: National Opinion Research Center.

Hoffer, T. B., S. Sederstrom, L. Selfa, V. Welch, M. Hess, S. Brown, S. Reyes, K. Webber, and I. Guzman-Barron. 2003. Doctorate Recipients from United States Universities: Summary Report 2002. Chicago: National Opinion Research Center.

Holden, C. 2004. Long hours aside, respondents say jobs offer 'as much fun as you can have.' *Science* 304(5678):1830-1837.

Holden, C. 2001. General contentment masks gender gap in first AAAS salary and job survey. *Science* 294(5541):396-411.

Hollenshead, C. S., B. Sullivan, G. C. Smith, L. August, and S. Hamilton. 2005. Work/family policies in higher education: Survey data and case studies of policy implementation. *New Directions for Higher Education* 130:41-65.

Hopkins, N. 2006. Diversification of a university faculty: Observations on hiring women faculty in the schools of science and engineering at MIT. *MIT Faculty Newsletter* 18(4):1, 16-23.

Hosek, S. D., A. G. Cox, B. Ghosh-Dastidar, A. Kofner, N. Ramphal, J. Scott, and S. H. Berry. 2005. Gender Differences in Major Federal External Grant Programs. RAND Technical Report. Santa Monica, CA: RAND Corporation.

Hyde, J. S., C. C. Hall, N. A. Fouad, G. P. Keita, M. E. Kite, N. F. Russo, and S. S. Brehm. 2002. Women in academe: Is the glass completely full? *American Psychologist* 57(12):1133-1134.

InterAcademy Council. 2006. *Women for Science*. Amsterdam: InterAcademy Council.

Ivie, R. 2004. Women physics and astronomy faculty. Presented to the Committee on Gender Differences in Careers of Science, Engineering, and Mathematics Faculty. The National Academies, Washington, DC. January 29.

Ivie, R., and K. N. Ray. 2005. *Women in Physics and Astronomy, 2005*. College Park, MD: American Institute of Physics.

Ivie, R., S. Guo, and A. Carr. 2005. *2004 Physics Academic Workforce Report*. College Park, MD: American Institute of Physics.

Ivie, R., K. Stowe, and K.Nies. 2003. *2002 Physics Academic Workforce Report*. College Park, MD: American Institute of Physics.

Jackson, J. 2004. The story is not in the numbers: Academic socialization and diversifying the faculty. *NWSA Journal* 16(1):172-185.

Jacobs, J. A., and S. Winslow. 2004. The academic life course, time pressures and gender inequality. *Community, Work & Family* 7(2):143-161.

Jacobs, J. A. 2004. The faculty time divide. *Sociological Forum* 19(1):3-27.

Jacobs, J. A. 1996. Gender inequality and higher education. *Annual Review of Sociology* 22: 153-185.

Johnson, G. 1999. Trends in the relative earnings of tenure track faculty: 1973-1995. Paper presented at the National Bureau of Economic Research, Cambridge, MA, April 21.

Johnson, G., and F. Stafford. 1974. The earnings and promotion of women faculty. *American Economic Review* 64(6):888-903.

Johnsrud, L., and V. Rosser. 2002. Faculty members' morale and their intention to leave. *Journal of Higher Education* 73(4):518-542.

Johnsrud, L., and R. Heck. 1998. Faculty worklife: Establishing benchmarks across groups. *Research in Higher Education* 39(5):539-555.

Kahn, S. 2002. The status of women in economics during the nineties: One step forward, two steps back. *CSWEP Newsletter* (Winter):5.

Kahn, S. 1995. Women in the economics profession. *The Journal of Economic Perspectives* 9(4): 193-206.

Kahn, S. 1993. Gender differences in academic career paths of economists. *American Economic Review* 83(2):52-56.

Keith, B., J. S. Layne, N. Babchuk, and K. Johnson. 2002. The context of scientific achievement: Sex status, organizational environments, and the timing of publication on scholarship outcomes. *Social Forces* 80(4):1253-1282.

Kierstad, D., P. D'Agnostino, and H. Dill. 1988. Sex role stereotyping of college professors: Bias in student ratings of instructors. *Journal of Educational Psychology* 80:342-344.

Kirby, K., R. Czujko, and P. Mulvey. 2001. The physics job market: From bear to bull in a decade. *Physics Today* 54(4):36-41.

Kirkman, E. E., J. W. Maxwell, and C. A. Rose. 2006. 2005 Annual survey of the mathematical sciences: Third report. *Notices of the AMS* 53(11):1345-1357.

Kirkman, E. E., J. W. Maxwell, and C. A. Rose. 2005. 2004 Annual survey of the mathematical sciences: Third Report. *Notices of the AMS* 52(8):871-883.

Kirkman, E. E., J. W. Maxwell, and C. A. Rose. 2004. 2003 Annual survey of the mathematical sciences: Third report. *Notices of the AMS* 51(8):901-912.

Kirkman, E. E., J. W. Maxwell, and K. R. Priestley. 2003. 2002 Annual survey of the mathematical sciences: Third report. *Notices of the AMS* 50(8):925-935.

Kite, M. E., N. F. Russo, S. S. Brehm, N. A. Fouad, C. C. Hall, J. S. Hyde, and G. P. Keita. 2001. Women psychologists in academe: Mixed progress, unwarranted complacency. *American Psychologist* 56(12):1080-1098.

Knapp, L. G., J. E. Kelly-Reid, R. W. Whitmore, S. Huh, L. Zhao, B. Levine, S. Ginder, J. Wang, and S. G. Broyles. 2005. Staff in Postsecondary Institutions, Fall 2003, and Salaries of Full-Time Instructional Faculty, 2003–04 (NCES 2005–155). U.S. Department of Education. Washington, DC: National Center for Education Statistics.

Koretz, G. 1997. Women storm the symphony. *Business Week* 3518(March 17):24.

Kreeger, K. 2001. *Women Chemists 2000.* Washington, DC: American Chemical Society.

Kreeger, K. 1999. Search committees: The long and winding road of academic hiring. *The Scientist* 13(22):24.

Krefting, L. A. 2003. Intertwined discourses of merit and gender: Evidence from academic employment in the USA. *Gender, Work and Organization* 10(2):260-278.

Kuck, V., C. H. Marzabadi, S. A. Nolan, and J. P. Buckner. 2004. Analysis by gender of the doctoral and postdoctoral institutions of faculty members at the top-fifty ranked chemistry departments. *Journal of Chemical Education* 81(3):356-363.

Kuck, V., J. P. Buckner, S. A. Nolan, and C. H. Marzabadi. 2003. The women's side of graduate school. *Chemical and Engineering News* 81(20):6-8.

Kuck, V. 2004. Transformation of the Top 10. *Chemical and Engineering News* 82(38):64-65.

Kuck, V. 2002. Women Physicists and Chemists are Making Slow Progress in Academe. *STATUS: A Report on Women in Astronomy* (January):1, 7, 19.

Kuck, V. 2001. Refuting the leaky pipeline hypothesis. *Chemical and Engineering News* 79(47):71-73.

Kuh, G. D., T. F. Nelson Laird, and P. D. Umbach, 2004. Aligning faculty activities and student behavior: Realizing the promise of greater expectations. *Liberal Education* (Fall):24-31.

Kuh, C. 2003. You've come a long way: Data on women doctoral scientists and engineers in research universities. Pp. 120-144 in *Equal Rites, Unequal Outcomes: Women in American Research Universities*, L. Hornig, ed. New York: Kluwer Academic Publishers.

Kulis, S. 1998. Organizational variations in women scientists' representation in academia. *Journal of Women and Minorities in Science and Engineering* 4(1):43-67.

Kulis, S. 1997. Gender segregation among college and university employees. *Sociology of Education* 70(2):151-173.

Kulis, S., D. Sicotte, and S. Collins. 2002. More than a pipeline problem: Labor supply constraints and gender stratification across academic science disciplines. *Research in Higher Education* 43(6):657-691.

Kulis, S., and D. Sicotte. 2002. Women scientists in academia: Geographically constrained to big cities, college clusters, or the coasts? *Research in Higher Education* 43(1):1-30.

Kulis, S., H. Shaw, and Y. Chong. 2000. External labor markets and the distribution of black scientists and engineers in academia. *The Journal of Higher Education* 71(2):187-222.

Kulis, S., Y. Chong, and H. Shaw. 1999. Discriminatory organizational contexts and black scientists on postsecondary faculties. *Research in Higher Education* 40(2):115-148.

Kulis, S., and K. A. Miller-Loessi. 1992. Organizational dynamics and gender equity. *Work and Occupations* 19(2):157-183.

Lane, N. 1999. Why are there so few women in science? *Nature Debates* (September 9). Available at http://www.nature.com/nature/debates.women/women_contents.html.

Lawler, A. 2005.. Summer's comments draw attention to gender, racial gaps. *Science* 307(5709): 492-493.

Lawler, A. 1999. Tenured women battle to make it less lonely at the top. *Science* 286(5443): 1272-12728.

Leo, J. 2002. Bogus bias at MIT. *U.S. News & World Report* 132(11):43.

Leslie, L., and R. Oaxaca. 1998. Women and minorities in higher education. Pp. 304-352 in *Higher Education: Handbook of Theory and Research*, vol. 13, J. C. Smart, ed. New York: Agathon Press.

Leslie, L., G. T. McClure, and R. Oaxaca. 1998. Women and minorities in science and engineering: A life sequence analysis. *The Journal of Higher Education* 69(3):239-276.

Lewis, D. R., and J. W. E. Becker, eds. 1979. *Academic Rewards in Higher Education*. Cambridge: Balinger Publishing Company.

Levin, I. P., R. M. Rouwenhorst, and H. M. Trisko. 2005. Separating gender biases in screening and selecting candidates for hiring and firing. *Social Behavior and Personality* 33(8):793-804.

Levin, S., and P. Stephan. 1998. Gender differences in the rewards to publishing in academe: Science in the 1970s. *Sex Roles* 38(11-12):1049-1064.

Levin, S., and P. Stephan. 1991. Research productivity over the life cycle: Evidence for academic scientists. *The American Economic Review* 81(1):114-132.

Lewin, A. Y., and L. Duchan. 1971. Women in academia. *Science* 173(4000):892-895.

Lindholm, J. 2003. Perceived organizational fit: Nurturing the minds, hearts, and personal ambitions of university faculty. *Review of Higher Education* 27(1):125-149.

Lipsey, M., and D. Wilson. 2000. *Practical Meta-Analysis.* Sage Publications on Applied Social Research Methods Series, vol. 49. Thousand Oaks, CA: Sage Publications.

Lively, K. 2000. Women in charge. *The Chronicle of Higher Education* 46(41):A33.

Loder, N. 2000. US science shocked by revelations of sexual discrimination. *Nature* 405(June 8):713-714.

Loeb, J. 2003. Hierarchical linear modeling in salary-equity studies. *New Directions for Institutional Research* 117:69-96.

Lomperis, A. M. T. 1990. Are women changing the nature of the academic profession? *The Journal of Higher Education* 61(6):643-677.

Long, J. R. 2002. Women still lag in academic ranks. *Chemical and Engineering News* 80(38): 110-111.

Long, J. R. 2000. Women chemists still rare in academia. *Chemical and Engineering News* 78(39):56.

Long, J. S. 1992. Measures of sex differences in scientific productivity. *Social Forces* 71(1): 159-178.

Long, J. S. 1990. The origins of sex differences in science. *Social Forces* 68(4):1297-1315.

Long, J. S., and M. F. Fox. 1995. Scientific careers: Universalism and particularism. *Annual Review of Sociology* 21:45-71.

Long, J. S., P. Allison, and R. McGinnis. 1993. Rank advancement in academic careers: Sex differences and the effects of productivity. *American Sociological Review* 58(5):703-722.

Long, J. S., P. Allison, and R. McGinnis. 1979. Entrance into the academic career. *American Sociological Review* 44(5):816-830.

Lubinski, D., C. P. Benbow, D. L. Shea, H. Eftekhari-Sanjani, and M. B. J. Malvorson. 2001. Men and women at promise for scientific excellence: Similarity not dissimilarity. *Psychological Science* 12(4):309-317.

Luna, A. L. 2006. Faculty salary equity cases: Combining statistics with the law. *Journal of Higher Education* 77(2):193-224.

Lundy, K. L. P., and B. D. Warme. 1990. Gender and career trajectory: The case of part-time faculty. *Studies in Higher Education* 15(2):207-223.

Luzzadder-Beach, S., and A. Macfarlane. 2000. The environment of gender and science: Status and perspectives of women and men in physical geography. *Professional Geographer* 52(3):407-424.

Manger, D. 1999. The graying professoriate. *The Chronicle of Higher Education*, 46(2):A18-A19.

Malcom, S. 1996. Science and diversity: A compelling national interest. *Science* 271(5257): 1817-1819.

Marasco, C. 2006. Women faculty gain little ground. *Chemical and Engineering News* 84(51): 58-59.

Marasco, C. 2005. Women faculty make little progress. *Chemical and Engineering News* 83(44): 38-39.

Marasco, C. 2004. No change in numbers of women faculty. *Chemical and Engineering News* 82(39):32-33.

Marasco, C. 2003. Numbers of women nudge up slightly. *Chemical and Engineering News* 81(43): 58-59.

Markus, G. 1979. *Analyzing Panel Data*. Newbury Park, CA: Sage Publications.

Marshke, R., S. Laursen, J. M. Nielsen, and P. Rankin. 2007. Demographic inertia revisited: An immodest proposal to achieve equitable gender representation among faculty in higher education. *The Journal of Higher Education* 78(1):1-26.

Martin, B. 2004. From industry to academia. *Nature* 429(May 20):324-325.

Marwell, G., R. Rosenfeld, and S. Spilerman. 1979. Geographic constraints on women's careers in academia. *Science* 205(4412):1225-1231.

Marzabadi, C. H., V. J. Kuck, S. A. Nolan, and J. P. Buckner, eds. 2006. *Are Women Achieving Equity in Chemistry? Dissolving Disparity, Catalyzing Change*. New York: Oxford University Press.

Mason, M. A., and M. Goulden. 2002. Do babies matter? The effect of family formation on the lifelong careers of academic men and women. *Academe* 88(6):21-27.

Mason, M. A., and M. Goulden. 2004a. Do babies matter (Part II)? Closing the baby gap. *Academe* 90(6):11-15.

Mason, M. A., and M. Goulden. 2004b. Marriage and baby blues: Redefining gender equity in the academy. *The Annals of the American Academy of Political and Social Science* 596(1):86-103.

Massachusetts Institute of Technology. 1999. A Study on the Status of Women Faculty in Science at MIT. Available at http://web.mit.edu/fnl/women/women.html.

McCormick, A. 2000. The 2000 Carnegie Classification: Background and Description (excerpt). Available at http://www.carnegiefoundation.org/Classification/CIHE2000/background.htm. Accessed January 27, 2005.

McCormick, A., ed. 2001. *The Carnegie Classification of Institutions of Higher Education, 2000 Edition*. Menlo Park, CA: The Carnegie Foundation for the Advancement of Teaching.

McDowell, J. M., L. D. Singell, Jr., and J. P. Ziliak. 1999. Cracks in the glass ceiling: Gender and promotion in the economics profession. *American Economic Review* 89(2):392-396.

McElrath, K. 1992. Gender, career disruption, and academic rewards. *The Journal of Higher Education* 63(3):269-281.

McLaughlin, G., and J. McLaughlin. 2003. Conducting a salary-equity study: A consultant's view. *New Directions for Institutional Research* 117:97-114.

McNeil, L., and M. Sher. 1999a. Dual-Science-Career Couples: Survey Results. Available at http://www.physics.wm.edu/dualcareer.html. Accessed May 13, 2004.

McNeil, L., and M. Sher. 1999b. The dual-career-couple problem. *Physics Today* 52(7):32-37.

Menges, R., and W. Exum. 1983. Barriers to the progress of women and minority faculty. *The Journal of Higher Education* 54(2):123-144.

Mervis, J. 2002. Can equality in sports be repeated in the lab? *Science* 298(5592):356.

Mervis, J. 2000. NSF in flux: NSF searches for right way to help women. *Science* 289(5478): 379-381.

Mielczarek, E. V. 2006. On the lack of women in academic science. *Science* 314(5799):592.

Milem, J. F, J. B. Berger, and E. L. Dey. 2000. Faculty time allocation: A study of change over twenty years. *The Journal of Higher Education* 71(4):454-475.

Miller, J. E., and C. Hollenshead. 2005. Gender, family, and flexibility—Why they're important in the academic workplace. *Change* 37(6):58-62.

Montelone, B. A., R. A. Dyer, D. J. Takemoto. 2003. A mentoring program for female and minority faculty members in the sciences and engineering: Effectiveness and status after 9 years. *Journal of Women and Minorities in Science and Engineering* 9(3/4):259-271.

Moody, J. 2004. Supporting women and minority faculty. *Academe* 90(1):47-52.

Muller, C. 2003. The underrepresentation of women in engineering and related sciences: Pursuing two complementary paths to parity. Pp. 119-126 in *Pan-Organizational Summit of the U.S. Science and Engineering Workforce: Meeting Summary,* M. A. Fox, ed. Washington, DC: The National Academies Press.

Nadis, S. 1999. Women scientists unite to battle cowboy culture. *Nature* 398(6726):361.

National Academy of Sciences, National Academy of Engineering, and Institute of Medicine. 2007. *Beyond Bias and Barriers: Fulfilling the Potential of Women in Academic Science and Engineering.* Washington, DC: The National Academies Press.

National Academy of Sciences, National Academy of Engineering, and Institute of Medicine. 2006. *Biological, Social, and Organizational Components of Success for Women in Academic Science and Engineering.* Washington, DC: The National Academies Press.

National Research Council. 2006. *To Recruit and Advance: Women Students and Faculty in U.S. Science and Engineering.* Washington, DC: The National Academies Press.

National Research Council. 2001a. *From Scarcity to Visibility: Gender Differences in the Careers of Doctoral Scientists and Engineers.* Washington DC: National Academy Press

National Research Council. 2001b. *Female Engineering Faculty at U.S. Institutions: A Data Profile.* Washington, DC: National Academy Press.

National Research Council. 1998. *Trends in the Early Careers of Life Scientists.* Washington, DC: National Academy Press.

National Research Council. Committee on Women in Science, Engineering and Medicine. http://sites. nationalacademies.org/pga/cwsem/PGA_045045.

National Science Board. 2006. *Science and Engineering Indicators 2006.* (vol. 1, NSB 06-01; vol. 2, NSB 06-01A). Arlington, VA: National Science Foundation.

National Science Board. 2004. *Science and Engineering Indicators 2004.* (vol. 1, NSB 04-1; vol. 2, NSB 04-1A). Arlington, VA: National Science Foundation.

National Science Board. 2002. *Science and Engineering Indicators – 2002.* (NSB-02-1). Arlington, VA: National Science Foundation.

National Science Foundation. 2007. *Women, Minorities, and Persons with Disabilities in Science and Engineering: 2007.* (NSF 07-309). Arlington, VA: NSF.

National Science Foundation. 2006. *Characteristics of Doctoral Scientists and Engineers in the United States: 2003.* (NSF 06-320). Project officer, John Tsapogas. Arlington, VA: NSF, Division of Science Resources Statistics.

National Science Foundation. 2004a. *Doctoral Scientists and Engineers: 2001 Profile Tables.* (NSF 04-312). Project officer, Kelly H. Kang. Arlington, VA: NSF, Division of Science Resources Statistics.

National Science Foundation. 2004b. *Gender Differences in the Careers of Academic Scientists and Engineers*. (NSF 04-323). Project officer, Alan I. Rapoport. Arlington, VA: NSF, Division of Science Resources Statistics.

National Science Foundation. 2004c. *Science and Engineering Doctorate Awards: 2003* (NSF 05-300). Project officer, Joan S. Burrelli. Arlington, VA: NSF, Division of Science Resources Statistics.

National Science Foundation. 2004d. *Women, Minorities, and Persons with Disabilities in Science and Engineering: 2004*. (NSF 04-317). Arlington, VA: NSF, Division of Science Resources Statistics. (updated May 2004).

National Science Foundation. 2003a. *Characteristics of Doctoral Scientists and Engineers in the United States: 2001*. (NSF 03-310). Project officer, Kelly H. Kang. Arlington, VA: NSF, Division of Science Resources Statistics.

National Science Foundation. 2003b. *Gender Differences in the Careers of Academic Scientists and Engineers: A Literature Review*. (NSF 03-322). Project director, Alan I. Rapoport. Arlington, VA: NSF, Division of Science Resources Statistics.

National Science Foundation. 2003c. *Graduate Students and Postdoctorates in Science and Engineering: Fall 2001*. (NSF 03-320). Project officer, Joan S. Burrelli. Arlington, VA: NSF, Division of Science Resources Statistics.

National Science Foundation. 2003d. *Science and Engineering Doctorate Awards: 2002*, (NSF 04-303). Project officer, Susan T. Hill. Arlington, VA: NSF, Division of Science Resources Statistics.

National Science Foundation. 2003e. *Women, Minorities, and Persons with Disabilities in Science and Engineering: 2002*. (NSF 03-312). Arlington, VA: NSF, Division of Science Resources Statistics.

National Science Foundation, 2002. *Science and Engineering Degrees: 1966-2000*. (NSF 02-327). Author, Susan T. Hill. Arlington, VA: NSF, Division of Science Resources Statistics.

National Science Foundation. 2000. *Women, Minorities, and Persons with Disabilities in Science and Engineering: 2000*. (NSF 00-327). Arlington, VA: NSF.

National Science Foundation. 1999. *Women, Minorities, and Persons with Disabilities in Science and Engineering: 1998*. (NSF 99-338). Arlington, VA: NSF.

Nelson, D., and Rogers, D. 2005. A National Analysis of Diversity in Science and Engineering Faculties at Research Universities. Available at http://cheminfo.chem.ou.edu/~djn/diversity/briefings/Diversity%20Report%20Final.pdf.

Nerad, M., and J. Cerny. 2002. Postdoctoral appointments and employment patterns of science and engineering doctoral recipients ten-plus years after Ph.D. completion: Selected results from the "Ph.D.s-Ten Years Later" study. *Communicator* 35(7):1-2, 10.

Nerad, M., and J. Cerny. 1999. Postdoctoral patterns, career advancement, and problems. *Science* 285(5433):1533-1535.

Nerad, M., and J. Cerny 1999. Widening the circle: Another look at women graduate students. *Communicator* 37(6):1-7.

Nerad, M., and C. L. Stewart. 1991. Assessing doctoral student experience: Gender and departmental culture. Paper presented at the 31st Annual Conference of the Association for Institutional Research, San Francisco, May 26-29.

Nettles, M. T., L. W. Perna, and E. M. Bradburn. 2000. Salary, Promotion, and Tenure Status of Minority and Women Faculty in U.S. Colleges and Universities (NCES 2000–173). U.S. Department of Education. Washington, DC: National Center for Education Statistics.

New Mexico State University. 2003. Space Allocation Survey. Available at http://www.advance.nmsu.edu/Documents/PDF/ann-rpt-03.pdf.

Nosengo, N. 2003. The quota conundrum. *Nature* 426(6963):211.

Oaxaca, R., and M. Ransom. 2002. Regression methods for correcting salary inequities between groups of academic employees. *New Directions for Institutional Re*search 115:91-103.

OER (Office of Extramural Research), National Institutes of Health. 2005. Sex/Gender in the Biomedical Science Workforce. Rockville, MD: NIH, Office of Extramural Research (October 7, 2005). Available at http://grants2.nih.gov/grants/policy/sex_gender/q_a.htm#q5.

O'Connell, S., and M. A. Holmes. 2004. Where are the women geoscience professors? *AWIS Magazine* 33(1):11-16.

Olsen, D., S. A. Maple, and F. K. Stage. 1995. Women and minority faculty job satisfaction: professional role interests, professional satisfactions, and institutional fit. *The Journal of Higher Education* 66(3):267-293.

Olson, K. 2002. Who gets promoted? Gender differences in science and engineering academia. *Journal of Women and Minorities in Science and Engineering* 8(3/4):347-362.

Olson, K. 1999. Despite increases, women and minorities still underrepresented in undergraduate and graduate S&E education. *NSF Data Brief,* January 15, (NSF 99-320).

Ongley, L. K., M. W. Bromley, and K. Osborne. 1998. Women geoscientists in academe: 1996-1997. *GSA Today* 8(11):12-13.

Packard, B. W. 2003. Web-based mentoring: Challenging traditional models to increase women's access. *Mentoring & Tutoring* 11(1):53-65.

Paludi, M. A., and L. A. Strayer. 1985. What's in an author's name? Differential evaluations of performance as a function of author's name. *Sex Roles* 12:353-361.

Park, S. 1996. Research, teaching, and service: Why shouldn't women's work count? *Journal of Higher Education* 67(1):46-84.

Parsad, B., and D. Glover. 2002. Tenure Status of Postsecondary Instructional Faculty and Staff: 1992–98 (NCES 2002–210). U.S. Department of Education. Washington, DC: National Center for Education Statistics.

Perkel, J. M. 2006. A lab startup. *The Scientist* 20(10):75.

Perna, L. 2005b. Sex differences in faculty tenure and promotion: The contribution of family ties. *Research in Higher Education* 46(3):277-307.

Perna, L. 2005. The relationship between family and employment outcomes. *New Directions for Higher Education* 130:5-23.

Perna, L. 2003a. Sex differences in faculty tenure and promotion: The contribution of family ties. Paper presented at the annual meeting of the Association for the Study of Higher Education, Portland, OR, November 14.

Perna, L. 2003b. Studying faculty salary equity: A review of theoretical and methodological approaches. Pp. 323-388 in *Higher Education: Handbook of Theory and Research,* vol. 18, J. C. Smart, ed. New York: Kluwer Academic Publishers.

Perna, L. 2003c. The status of women and minorities among community college faculty. *Research in Higher Education* 44(2):205-240.

Perna, L. 2002. Sex differences in the supplemental earnings of college and university faculty. *Research in Higher Education* 43(1):31-58.

Perna, L. 2001a. Sex and race differences in faculty promotion. *Research in Higher Education* 42(5):541-567.

Perna, L. 2001b. Sex differences in faculty salaries: A cohort analysis. *The Review of Higher Education* 24(3):283-807.

Perna, L. 2001c. The relationship between family responsibilities and employment status among college and university faculty. *The Journal of Higher Education* 72(5):584-611.

Persell, C. H. 1983. Gender, rewards and research in education. *Psychology of Women Quarterly* 8(33):46.

Phillips, S., and M. Garner. 2006. Where "hello baby" doesn't mean "bye bye tenure." *Times Higher Education Supplement* 1732(March 3):18-19.

Porter, S., and P. Umbach. 2001. Analyzing faculty workload data using multilevel modeling. *Research in Higher Education* 42(2):171-196.

Preston, A. E. 2004. Plugging the leaks in the scientific workforce. *Issues in Science and Technology* 20(4):69-74.

Preston, A. E. 1994. Why have all the women gone? A study of exit of women from the science and engineering professions. *American Economic Review* 84:1446-1462.

Qamar uz, M. 2004. *Review of the Academic Evidence on the Relationship between Teaching and Research in Higher Education.* DfES Research Report no. 506. London: U. K. Department for Education and Skills.

Quinn, K., S. E. Lange, and S. G. Olswang. 2004. Family-friendly policies and the research university. *Academe* 90(6):32-34.

Radetsky, P. 1994. The modern postdoc: Prepping for the job market. *Science* 265(5180): 1909-1910.

Rankin III, S. 2004. Studying gender differences among science, engineering, and mathematics faculty. Presented to the Committee on Gender Differences in Careers of Science, Engineering, and Mathematics Faculty. The National Academies. Washington, DC. January 29.

Ransom, M., and S. Megdal. 1993. Sex differences in the academic labor market in the affirmative action era. *Economics of Education Review* 12(1):21-43.

Rausch, D. K., B. P. Ortiz, R. A. Douthitt, and L. L. Reed. 1989. The academic revolving door: Why do women get caught? *CUPA Journal* 40(1):1-16.

Rayman, P., and B. Brett. 1995. Women science majors: What makes a difference in persistence after graduation? *The Journal of Higher Education* 66(4):388-414.

Regets, M. 1998. Has the use of postdocs changed? *NSF Issue Brief*, December 2, (NSF 99-310).

Reis, R. 1999. The right start-up package for beginning science professors. *Chronicle of Higher Education*, August 27. Available at http://chronicle.com/article/The-Right-Start-Up-Package-/45611.

Reskin, B. 1976. Sex differences in status attainment in science: The case of the postdoctoral fellowship. *American Sociological Review* 41(4):597-612.

Reskin, B., and D. D. Bielby. 2005. A sociological perspective on gender and career outcomes. *Journal of Economic Perspectives* 19(1):71-86.

Rhoads, S. E., and C. H. Rhoads. 2004. Gender roles, infant care and gender neutral post-birth paid leave policies in America. Paper presented at the Annual Meeting of the Midwest Political Science Association, Chicago, April 16.

Riger S., J. Stokes, S. Raja, and M. Sullivan. 1997. Measuring perceptions of the work environment for female faculty. *The Review of Higher Education* 21:63-78.

Rimer, S. "For Women in Science, Slow Progress in Academia," *New York Times*, April 15, 2005.

Rivard, N. 2003. Who's running the show? *University Business* 6(2):11.

Roberts, H. 1993. Can statistics tell us what we do not want to hear? The case of complex salary structures: Comment. *Statistical Science* 8(2):171-176.

Robst, J., J. Keil, and D. Russo. 1998. The effect of gender composition of faculty on student retention. *Economics of Education Review* 17(4):429-439.

Rosenfeld, R., and J. A. Jones. 1987. Patterns and effects of geographic mobility for academic women and men. *The Journal of Higher Education* 58(5):493-515.

Rosser, S. 2006. Senior women scientists overlooked and understudied. *Journal of Women and Minorities in Science and Engineering* 12(4):275-293.

Rosser, S. 2004. Using POWRE to ADVANCE: Institutional barriers identified by women scientists and engineers. *NWSA Journal* 16(1):50-78.

Rosser, S. 2003. Attracting and retaining women in science and engineering. *Academe* 89(4):24-29.

Rosser, S., and J. Z. Daniels. 2004. Widening paths to success, improving the environment, and moving toward lessons learned from the experiences of POWRE and CBL awardees. *Journal of Women and Minorities in Science and Engineering* 10(2):131-148.

Rosser, S., J. Z. Daniels, and L. Wu. 2006. Institutional factors contributing to the dearth of women STEM faculty: Classification and status matter; location doesn't. *Journal of Women and Minorities in Science and Engineering* 12(1):79-93.

Rosser, S., and E. O. Lane. 2002. A history of funding for women's programs at the National Science Foundation: From individual POWRE approaches to the ADVANCE of institutional approaches. *Journal of Women and Minorities in Science and Engineering* 8(3/4):327-346.

Rosser, S., and E. O. Lane. 2002. Key barriers for academic institutions seeking to retain female scientists and engineers: Family-unfriendly policies, low numbers, stereotypes, and harassment. *Journal of Women and Minorities in Science and Engineering* 8(2):161-189.

Rosser, V. 2005. Measuring the change in faculty perceptions over time: An examination of their worklife and satisfaction. *Research in Higher Education* 46(1):81-107.

Rosser, V. 2004. Faculty members' intentions to leave: A national study on their worklife and satisfaction. *Research in Higher Education* 45(3):285-309.

Rothblum, E. D. 1988. Leaving the ivory tower: Factors contributing to women's voluntary resignation from academia. *Frontiers: A Journal of Women Studies* 10(2):14-17.

Rubin, A., and D. M. Powell. 1987. Gender and publication rates: A reassessment with population data. *Social Work* 32(4):317-320.

Sabatier, M., M. Carrere, and V. Mangematin. 2006. Profiles of academic activities and careers: Does gender matter? An analysis based on French life scientist CVs. *Journal of Technology Transfer* 31(3):311-324.

Sadrozinski, R., M. Nerad, and J. Cerny. 2003. *PhDs in Art History—Over a Decade Later: A National Career Path Study of Art Historians*. Seattle: CIRGE, University of Washington.

Sakai, A., and M. Lane. 1996. National Science Foundation funding patterns of women and minorities in biology. *Bioscience* 46(8):621-626.

Sands, R. G., L. A. Parson, and J. Duane. 1991. Faculty mentoring faculty in a public university. *The Journal of Higher Education* 62(2):174-193.

Saunders, N. 2004. Reach out and cuddle up to another discipline. *The Times Higher Education Supplement* 1621(January 2):21.

Sax, L., L. S. Hagedorn, A. Marisol, and F. A. DiCrisi. 2002. Faculty research productivity: Exploring the role of gender and family-related factors. *Research in Higher Education* 43(4):423-446.

Schneider, A. 2000. Female scientists turn their backs on jobs at research universities. *Chronicle of Higher Education* 46(50):A12-A14.

Schneider, A. 1998. Why don't women publish as much as men? *Chronicle of Higher Education* 45(3):A14-A16.

Sears, A. L. W. 2003. Image problems deplete the number of women in academic applicant pools. *Journal of Women and Minorities in Science and Engineering* 9(2):169-181.

Selvin, P. 1992. Profile of a field: Mathematics. *Science* 255(5050):1382-1383.

Sharpe, N. R., and G. Sonnert. 1999. Women mathematics faculty: Recent trends in academic rank and institutional representation. *Journal of Women and Minorities in Science and Engineering* 5(3):207-217.

Sidanius, J., and M. Crane. 1989. Job evaluation and gender: The case of university faculty. *Journal of Applied Social Psychology* 19:174-197.

Singer, M. 2004. The evolution of postdocs. *Science* 306(5694):232.

Slade, D. 1999. Dual-career couples. *Chemical and Engineering News* 77(46):61-63.

Sonnert, G., with G. Holton. 1995. *Who Succeeds in Science? The Gender Dimension*. New Brunswick, NJ: Rutgers University Press.

Sonnert, G., with G. Holton. 1995. *Gender Differences in Science Careers: The Project Access Study*. New Brunswick, NJ: Rutgers University Press.

Stacey, A. M. 2003. *Report on the University of California, Berkeley Faculty Climate Survey, 2003*. Berkeley, CA: University of California, Berkeley.

Stack, S. 2004. Gender, children and research productivity. *Research in Higher Education* 45(8): 891-920.

Stake, J. E., Walker, E. F., and Speno, M. V. 1981. The relationship of sex and academic performance to quality of recommendations for graduate school. *Psychology of Women Quarterly* 5:515-522.

Steinpreis, R., K. Anders, and D. Ritzke. 1999. The impact of gender on the review of the curricula vitae of job applicants and tenure candidates: A national empirical study. *Sex Roles* 41(7/8):509-528.

Strober, M., and A. Quester. 1977. The earnings and promotion of women faculty: Comment. *American Economic Review* 67(2):207-213.

Sullivan, B., C. Hollenshead, and G. Smith. 2004. Developing and implementing work-family policies for faculty. *Academe* 90(6):24-27.

Swim, J., E. Borgida, G. Maruyama, and D. G. Myers. 1989. Joan McKay versus John McKay: Do gender stereotypes bias evaluations? *Psychological Bulletin* 105(3):409-429.

Tang, J. 2003. Women succeeding in science in the twentieth century. *Sociological Forum* 18(2): 325-342.

Tang, J. 1997. The glass ceiling in science and engineering. *Journal of Socio-Economics* 26(4): 383-406.

Tatro, C. N. 1995. Gender effects of student evaluations of faculty. *Journal of Research and Development in Education* 28:169-173.

Tesch, B. J., H. M. Wood, A. L. Helwig, and A. B. Nattinger. 1995. Promotion of women physicians in academic medicine: Glass ceiling or sticky floor? *JAMA* 273(13):1022-1025.

Thornton, S. 2005. Implementing flexible tenure clock policies. *New Directions in Higher Education* 130:81-90.

Thorton, S. 2004. Where—not when—should you have a baby? *Chronicle of Higher Education,* 51(7):B12.

Tilghman, S. 2004. Ensuring the Future Participation of Women in Science, Mathematics, and Engineering. Pp. 7-12 in *The Markey Scholars Conference: Proceedings.* Washington, DC: The National Academies Press.

Tolbert, P.S. 1995. The effects of gender composition in academic departments on faculty turnover. *Industrial and Labor Relations Review* 48(3):562-579.

Toutkoushian, R., and V. M. Conley. 2005. Progress for women in academe, yet inequities persist: Evidence from NSOPF:99. *Research in Higher Education* 46(1):1-28.

Toutkoushian, R., and M. Bellas. 2003. The effects of part-time employment and gender on faculty earnings and satisfaction. *The Journal of Higher Education* 74(2):172-195.

Toutkoushian, R., S. R. Porter, C. Danielson, and P. R. Hollis. 2003. Using publication counts to measure an institution's research productivity. *Research in Higher Education* 44(2):121-148.

Toutkoushian, R., and E. Hoffman. 2002. Alternatives for measuring the unexplained wage gap. *New Directions for Institutional Research* 115:71-90.

Toutkoushian, R. 1998a. Racial and marital status differences in faculty pay. *The Journal of Higher Education* 69(5):513-541.

Toutkoushian, R. 1998b. Sex matters less for younger faculty: Evidence of disaggregate pay disparities from the 1988 and 1993 NCES surveys. *Economics of Education Review* 17(1):55-71.

Trautner, J. J., K. C. Chou, J. K. Yates, and J. Stalnaker. 1996. Women faculty in engineering: Changing the academic climate. *Journal of Engineering Education,* 85:45-51.

Trix, F., and C. Psenka. 2003. Exploring the color of glass: Letters of recommendation for female and male medical faculty. *Discourse & Society* 14(2):191-220.

Trower, C. A. 2002. Women without tenure, Part 3: Why they leave. *Science* (March 22). Available at http://sciencecareers.sciencemag.org/career_magazine/previous issues/aticles/2002_0_3_22/noDOI.8430223210060551899.

Trower, C. A. 2002. Women without tenure, Part II: The gender sieve. *Science* (January 25). Available at http://sciencecareers.sciencemag.org/career_magazine/previous issues/aticles/2002_0_1_25/noDOI.7900867231599505905.

Trower, C. A. 2001. Women without tenure, Part 1. *Science* (September 14). Available at http://sciencecareers.sciencemag.org/career_magazine/previous issues/aticles/2002_0_9_14/noDOI.10565480637185635938.

Trower, C. A., and J. L. Bleak, 2004. Study of New Scholars. *Gender: Statistical Report [Universities]*. Cambridge, MA: Harvard University, Graduate School of Education.

Trower, C. A., and R. Chait. 2002. Faculty diversity. *Harvard Magazine* (March-April):33-37, 98.

Tuckman, H. P., and R. P. Hagemann. 1976. An analysis of the reward structure in two disciplines. *Journal of Higher Education* 47(4):447-464.

Turner, C. S. V., S. L. Myers, Jr., and J. W. Cresswell. 1999. Exploring underrepresentation: The case of faculty of color in the Midwest. *The Journal of Higher Education* 70(1):27-59.

Umbach, P. D. 2007. Gender equity in the academic labor market: An analysis of academic disciplines. *Research in Higher Education* 48(2):169-192.

University of Pennsylvania Almanac, Almanac Supplement. 2001. Vol. 48. no. 14. Available at http://www.upenn.edu/almanac/v48/n14/GenderEquity.html. Accessed April 22, 2009.

Urry, M. 2000. The status of women in astronomy. *STATUS: A Report on Women in Astronomy* (June):1-4, 7.

Urry, M., and V. Kuck. 2002. Addendum: 'Yields' and 'parity indices' for top astronomical institutions. *STATUS: A Report on Women in Astronomy* (January):8-9.

U.S. Department of Education, 2003. *The Condition of Education 2003*. (NCES 2003-067). U.S. Department of Education. Washington, DC: National Center for Education Statistics.

U.S. Department of Education. 2002. 1993 National Study of Postsecondary Faculty (NSOPF:93), Part-time Instructional Faculty and Staff: Who They Are, What They Do, and What They Think. (NCES 2002-163). By Valerie M. Conley and David W. Leslie. Project officer: Linda J. Zimbler. Washington, DC: National Center for Education Statistics.

U.S. Department of Education. 2002. Supplemental Table Update to 1993 National Study of Postsecondary Faculty (NSOPF:93), Part-time Instructional Faculty and Staff: Who They Are, What They Do, and What They Think. (NCES 2002-163). By Valerie M. Conley and David W. Leslie. Project officer: Linda J. Zimbler. Washington, DC: National Center for Education Statistics.

U.S. Department of Education. 2001. *Competing Choices: Men's and Women's Paths After Earning a Bachelor's Degree. (*NCES 2001–154). By Michael S. Clune, Anne-Marie Nuñez, and Susan P. Choy. Project officer: C. Dennis Carroll. Washington, DC: National Center for Education Statistics.

U.S. Department of Education. 2001. Postsecondary Institutions in the United States: Fall 2000 and Degrees and Other Awards Conferred: 1999–2000. (NCES 2002–156). By Laura G. Knapp, et. al. Project officer: Susan G. Broyles. Washington, DC: National Center for Education Statistics.

U.S. Department of Education. 2000. Entry and Persistence of Women and Minorities in College Science and Engineering Education. (NCES-2000-601). Washington, DC: National Center for Education Statistics.

U.S. Department of Education. 2000. Financial Aid Profile of Graduate Students in Science and Engineering. (Working Paper No. 2000-11). By Lawrence K. Kojaku. Project officer, Dennis Carroll. Washington, DC: National Center for Education Statistics.

Valian, V. 2004. Beyond gender schemas: Improving the advancement of women in academia. *NWSA Journal* 16(1):207-220.

Valian, V. 1998. *Why So Slow? The Advancement of Women*. Cambridge, MA: Massachusetts Institute of Technology Press.

Valian, V., and A. Stewart. 2005. Letters to the editor. *The Chronicle of Higher Education* 51(20): A47.

Van Anders, S. 2004. Why the academic pipeline leaks: Fewer men than women perceive barriers to becoming professors. *Sex Roles* 51(9/10):511-521.

Van Ommeren, J., R. E. de Vries, G. Russo, and M. Van Ommeren. 2005. Context in selection of men and women in hiring decisions: Gender composition of the applicant pool. *Psychological Reports* 96(2):349-360.

Van Ommersen, C. A. 2005. No talent left behind: Attracting and retaining a diverse faculty. *Change* 37(6):27-31.

Vander Putter, J., and L. Wimsatt. 1998. Faculty and departure: An international test of two models. Paper presented at the 23rd Annual Meeting of the Association for the Study of Higher Education, Miami, FL, November 5-8.

Vardi, M., T. Finin, and T. Henderson. 2003. 2001-2002 Taulbee Survey: Survey results show better balance in supply and demand. *Computing Research News* 15(2):6-13.

Vitulli, M., and M. Flahive. 1997. Are women getting all the jobs? *Notices of the American Mathematical Society* 44(3):338-339.

Volk, C. S., S. Slaughter, and S. L. Thomas. 2001. Models of institutional resource allocation. *The Journal of Higher Education* 72(4):387-413.

Ward, K., and L. E. Wolf-Wendel. 2005. Work and family perspectives from research university faculty. *New Directions for Higher Education* 130:67-80.

Ward, K., and L. E. Wolf-Wendel. 2004b. Fear factor: How safe is it to make time for family? *Academe* 90(6):28-31.

Ward, K., and L. E. Wolf-Wendel. 2004. Academic motherhood: Managing complex roles in research universities. *The Review of Higher Education* 27(2):233-257.

Weiss, Y., and L. Lillard. 1982. Output variability, academic labor contracts, and waiting times for promotion. *Research in Labor Economics* 5:157-188.

Wells, C. 2003. When tenure isn't enough. *The Chronicle of Higher Education* 50(10):C3.

Wenneras, C., and A.Wold. 1997. Nepotism and sexism in peer-review. *Nature* 387(6631):341-343.

West, M. S., and J. W. Curtis. 2006. *AAUP Faculty Gender Equity Indicators 2006.* Washington, DC: American Association of University Professors.

White, L. 2003. Deconstructing the public-private dichotomy in higher education. *Change* 35(3):49-54.

Williams, J. 2005. The glass ceiling and the maternal wall in academia. *New Directions for Higher Education* 130, 91-105.

Williams, J. 2004. Hitting the maternal wall. *Academe* 90(6):16-20.

Williams, J. 2003. The subtle side of discrimination. *Chronicle of Higher Education* 49(32):C5.

Williams, J. 2000. What stymies women's academic careers? It's personal. *Chronicle of Higher Education* 47(16):B10.

Wilson, R. 2005. Rigid tenure system hurts young professors and women, university officials say. *Chronicle of Higher Education* 52(7):A12.

Wilson, R. 2004. Where the elite teach, it's still a man's world. *Chronicle of Higher Education* 51(15):A8-14.

Wilson, R. 2003. Duke and Princeton will spend more to make female professors happy. *Chronicle of Higher Education* 50(7):A12.

Wilson, R. 2003. How babies alter careers for academics. *Chronicle of Higher Education* 50(15):1.

Wilson, R. 2002. Stacking the deck for minority candidates? *Chronicle of Higher Education* 48(44):A10-A12.

Wilson, R. 2001. The backlash against hiring couples. *Chronicle of Higher Education* 47(31):A16-A18.

Wilson, R. 2000. Female scholars suggest slowing tenure clock. *Chronicle of Higher Education* 46(20):A18.

Wilson, R. 1999. An MIT professor's suspicion of bias leads to a new movement for academic women. *Chronicle of Higher Education* 46(15):A16.

Winkler, J. 2000. Faculty reappointment, tenure, and promotion: Barriers for women. *Professional Geographer* 52(4):737-750.

Winkler, J. A., D. Tucker, and A. K. Smith. 1996. Salaries and advancement of women faculty in atmospheric science: Some reasons for concern. *Bulletin of the American Meteorological Society* 77(3):473-490.

Wolf-Wendel, L. E., and K. Ward. 2006. Academic life and motherhood: Variations by institutional type. *Higher Education* 52(3):487-521.

Wolf-Wendel, L. E., S. B. Twombly, and S. Rice. 2003. *The Two-Body Problem: Dual-Career-Couple Hiring Practices in Higher Education*. Baltimore: The Johns Hopkins University Press.

Wolf-Wendel, L., S. B. Twombly, and S. Rice. 2000. Dual-career couples: Keeping them together. *Journal of Higher Education* 71(3):291-321.

Women in Science & Engineering Leadership Institute (WISELI), University of Wisconsin-Madison. Reviewing Applicants: Research on Bias and Assumptions. Available at http://wiseli.engr.wisc. edu/initiatives/hiring/Bias.pdf. Accessed April 20, 2005.

Workshop on Building Strong Academic Chemistry Departments Through Gender Equity, complete report. 2006. (Arlington, VA, January 29-21, 2006). Available at http://www.chem.harvard.edu/ groups/friend/GenderEquityWorkshop/GenderEquity.pdf. Accessed December 18, 2006.

Wright A. L., L. A. Schwindt, T. L. Bassford. V. F. Reyna, C. M. Shisslak, P. A. St. Germain, and K. L. Reed. 2003. Gender differences in academic advancement: Patterns, causes, and potential solutions in one U.S. college of medicine. *Academic Medicine* 78(5):500-508.

Wyden, R. 2003. Title IX and women in academics. *Computing Research News* 15(4):1, 8.

Wyden, R. 2002. Statement. Presented at the hearing on Title IX and Science, U.S. Senate Committee on Commerce, Science and Transportation, Washington, DC, October 3.

Xie, Y., and K. Shauman. 2003. *Women in Science: Career Processes and Outcomes*. Cambridge: Harvard University Press.

Yamagata, H. 2002. Trends in faculty attrition at U.S. medical schools, 1980-1999. *Analysis in Brief* (Association of American Medical Colleges), 2(2):1-2.

Yarnell, A. 2003. Grad school: Does it matter where you go? *Chemical and Engineering News* 81(39):42, 44.

Yoder, J., P. Crumpton, and J. Zipp. 1989. The power of numbers in influencing hiring decisions. *Gender and Society* 3(2):269-276.

Yoest, C. 2004. Parental Leave in Academia. Available at http://faculty.virginia.edu/familyandtenure/ institutional%20report.pdf. Accessed April 24, 2009.

Zare, R. N. 2006. Sex, lies, and Title IX. *Chemical & Engineering News* 84(20):46-49.

Zhou, Y., and J. F. Volkwein. 2004. Examining the influences on faculty departure intentions: A comparison of tenured versus nontenured faculty at research universities using NSOPF-99. *Research in Higher Education* 45(2):139-176.

Zuckerman, H. 2001. The careers of men and women scientists: Gender differences in career attainments. Pp. 69-78 in *Women, Science, and Technology: A Reader in Feminist Studies*, M. Wyer, M. Barbercheck, D. Ozturk, H. O. Ozturk, and M. Wayne, eds. New York: Routledge.

Zuckerman, H. 1991. The careers of men and women scientists: A review of current research. In *The Outer Circle: Women's Position in the Scientific Community*, H. Zuckerman, J. Cole, and J. Bruer, eds. New York: W. W. Norton.

Zweben, S. 2005. Record Ph.D. production on the horizon; undergraduate enrollments continue in decline. *Computing Research News*. May. Pp. 7-15.

Index

A

Academic hiring. *See also* Applications;
   Interviews; Offers; Recruitment
   data on, 42-43, 44-48, 50-54
   equity in, 4, 27-28, 36, 63, 68, 274
   faculty perspective, 61
   key findings, 5-8, 64-69, 154-158
   process, 5, 40-43, 275
   productivity and, 88
   selection process, 41, 42, 50-56
   recommendations, 164-165
   of spouses, 274, 279
   statistical analyses, 43, 48-50, 54-56, 57-59
   survey questionnaire, 193-196
   target-of-opportunity positions, 44, 46, 64
   trends. 16, 148-149, 154
   women's involvement in, 5, 50, 55, 154,
      157, 279
Academics. *See* Assistant professors; Associate
   professors; Faculty; Full professors
ADVANCE program, 16, 18
Age factors, 36, 270-271, 326
Agricultural sciences, 2, 6, 20, 25 n.14, 32, 35,
   37 n.19, 60, 155, 249, 254, 255, 262
American Association for the Advancement of
   Science, 32, 103
American Association of University Professors,
   32, 36 n.16, 37, 103, 269

American Chemical Society, 32, 269, 277, 280,
   281
American Institute of Physics, 28, 32, 181, 182,
   280
American Mathematical Society, 32
American Society for Engineering Education,
   32
Antidiscrimination laws, 17-18
Applications for faculty positions
   climate perceptions and, 277-278, 282
   decision to apply, 41
   descriptive data, 44-48
   discipline and, 43, 46, 47, 48, 49-50, 65, 66,
      68, 154, 156, 277, 284-286
   doctoral pool relative to, 21, 23, 28, 46-48,
      61, 65, 66, 68, 154, 156, 164, 277-278
   faculty composition and, 278-279
   family-friendly policies and, 49, 50, 62, 63,
      66, 279, 282
   geographic mobility constraints, 21, 271,
      280-281
   by institution type, 47, 49, 50, 66, 279, 284,
      285, 286
   institutional policies for increasing diversity,
      8, 61-64, 66, 274
   key findings, 65-66, 154-156
   male-only pools, 65, 66, 154, 284
   pool of female applicants, 5, 7, 8, 23, 28,
      44, 154, 156

B

U